W
REALLY
WHO?

WHO IS REALLY WHO?

The Comprehensive guide to DNA paternity testing

Avi Lasarow

JOHN BLAKE

Published by John Blake Publishing Ltd,
3, Bramber Court, 2 Bramber Road,
London W14 9PB, England

www.blake.co.uk

First published in hardback in 2006

ISBN 1 84454 226 2

British Library Cataloguing-in-Publication Data:

A catalogue record for this book is available from the British Library.

Design by www.envydesign.co.uk

Printed in Great Britain by Bookmarque Ltd, Croydon, Surrey

1 3 5 7 9 10 8 6 4 2

Papers used by John Blake Publishing are natural, recyclable
products made from wood grown in sustainable forests.
The manufacturing processes conform to the environmental
regulations of the country of origin.

Every attempt has been made to contact the relevant
copyright-holders, but some were unobtainable. We would be
grateful if the appropriate people could contact us.

In memory of the late Bernard Leonard
Lasarow, 1938–2004
A fantastic father and a wonderful dad.

'The best parent is both parents.'

CONTENTS

INTRODUCTION

The first time I ever heard of paternity testing was, probably like many people, on a late-night American TV show. The thought that it would ever become relevant either to me or my friends was too remote to contemplate. However, three years ago I had cause to investigate the market, and I was horrified to find that all the available websites offering 'home kits' were hosted by non-UK-based companies, generally American outfits contactable only via answering-machine services and post office box numbers. When I investigated the situation further and approached a GP, the only help I got was a referral to a few website addresses, all of which I had already visited.

When I contacted the foreign websites, I found

them to be devoid of emotional support and clinical in the extreme. I observed that the companies didn't ask about motives and expectations, or warn of any possible reaction to the life-changing news they would supply. I have no doubt that the results that they provided were accurate and wouldn't want to cast aspersions on any such company, but to my mind the services they offered lacked sensitivity. Such companies seemed to be more concerned about each client's financial contribution than his or her emotional wellbeing, and ensuring that each client's credit-card details were correct was judged to be more important than helping him or her deal with any unfavourable news with which he or she might ultimately be presented.

It also concerned me that these companies were not contactable directly by telephone, and nor did they have offices for clients to visit, so any client with concerns about the service with which he or she was provided or about the results supplied had no recourse. UK law doesn't govern these companies, and their practices are not subject to the UK's Trading Standards Act. In theory, any test results they supplied could have been fictitious.

A few months after I started investigating DNA (deoxyribonucleic acid, the fundamental compound involved in genetics) testing, I was made redundant from my job. Having spent several years working in banks, this seemed to be the perfect opportunity to change the direction of my career. After a great deal of thought and research, I put together a company offering a new concept in DNA-based relationship testing.

I felt that we should place more importance upon the emotional welfare of the individual who sought out our help than other companies that were offering the service at that time. Whether a child is looking for a parent or a parent is attempting to claim a child, all clients need to be aware of the consequences of such investigations. Having witnessed the difficulties that others have experienced in their searches for answers, it has become a mission of mine to raise public awareness of DNA testing and the emotional consequences of such a process

I researched laboratories, looking for respectable institutions in which I could be confident that the samples would be kept secure and the results accurate. Ultimately I made sure that all of the establishments used were accredited and had a history of performing to the highest standards. I felt it was important that the process was as simple and painless as possible, and that the results were made available to our clients at the earliest achievable moment.

I decided that it was necessary to ensure that whoever approached us felt that their enquiries would be treated confidentially. To this end, I designed privacy policies ensuring that we keep no client information for marketing purposes and treat all correspondence with the utmost discretion.

Before the company I founded agrees to do a test, we make certain that the customer has thought about the possible results and understands the implication of each. We make sure that they're as prepared as possible for any eventuality. To this end, we work closely with UK-based charities such as Families Need Fathers (FNF), supporting both parents and children through the testing process.

Most men believe that the likelihood of them ever becoming involved in paternity testing is negligible, yet recent revelations have proved that no one is immune to mistakes. Anyone, from any background, whether a cabinet minister or a film star, might require these types of services. Contrary to popular belief, most of the cases that we deal with aren't presented by men attempting to dodge child-support responsibilities; most are simply looking for peace of mind. The following is a classic example of the kind of situation that we deal with on a daily basis:

'Donna' had split up with 'Kevin' just three weeks before meeting an old flame, 'Andrew'. Two months into her new relationship, she found out that she was pregnant. Donna was more than sure that the baby was Andrew's, and so was Andrew. However, 'Kevin' decided that he would attempt to gain access to the child through the courts, claiming that it was his. We tested both Andrew and the baby, and were delighted to confirm that the child was indeed Andrew's. Both parents were ecstatic to the point of screaming with delight, and we were happy to celebrate with them.

The process of DNA testing is very much a double-edged sword. There are sad instances when fathers find out that a child they adore is not legally theirs, but equally there are those when families are reunited after their DNA has been tested during the immigration process

I feel that it's essential to raise the public's awareness of the consequences of having a DNA test carried out and the effects that it can have on individuals and families. I also feel that it's important that fathers have the right to know whether or not the child in whom they are investing not only their money but also their emotions is in fact theirs. Likewise, the children have a right to know who they are. However, all clients need to be alert to the potential emotional cost.

Since I began my career in DNA-based paternity testing, I've encountered people from all age groups seeking an answer to a question that has been gnawing at them for years: are they really related to their brothers or sisters? Does that long-standing family 'joke' bear any proximity to the truth? The emotional damage I've witnessed over the last few years caused by dubious paternity far outweighs any criticism against the services that companies such as ours provide.

Because of the need we see every day at my company, I felt that it was time to produce a guide that would steer people through the legal and practical process, as well as addressing the intrinsic emotional consequences. In the following pages, therefore, I've identified the kinds of samples required for the test and explained how the process works. If you're in a position where you're thinking of taking such a test, I ask you to do two things: assess your reasons for doing so, and research the support that's currently available in the UK in the form of a large number of charities and government bodies who can assist in such matters.

I sincerely hope that the following pages will remove any confusion about the process of DNA-based paternity testing and will enable any interested reader to make an informed decision concerning his or her next step into finding the truth behind his or her family history.

1

WHAT IS DNA PATERNITY TESTING?

THE HISTORY OF DNA TESTING

Until relatively recently, methods used to identify the father of a child were basic and of limited accuracy. Before the introduction of blood testing in the 1920s, the only method of identifying the father of a child born outside marriage was through the testimony of the mother, which was taken by magistrates. The expectation was that the father, once identified in this way, was to contribute support to the parish for the upbringing of the child. This procedure obviously failed to take into account the possibility that the mother was mistaken or, perhaps, had volunteered information that was deliberately inaccurate. The assumption was that any child conceived within marriage was legitimate, provided that the father was resident at the time of conception and was physically capable of fathering a child. In fact, even with the advent of DNA testing, the first case in England of a man contesting his wife's claim for £300,000 in child maintenance in a divorce settlement, after establishing that the children had been fathered by another man, was brought as recently as 2002.

Science entered the paternity arena in the 1920s with the introduction of blood typing, based on the ABO blood-group system. This is not, however, an accurate method for determining paternity, as the results simply eliminate (ie exclude) only 30 per cent of the entire male population from being the possible father. (When the term *exclusion* is used in terms of paternity testing, it refers to an individual *not* having a biological relationship with another. Conversely, the term *inclusion* refers to a person having such a relationship.) Essentially, the test is a means of eliminating the possibility of any individual sharing a biological relationship with another individual. Such a method provides an indication of

2

parentage that can be used for establishing paternity but is not as accurate as today's methods of testing.

The accuracy of paternity testing advanced during the 1930s with the introduction of serological testing, which is a form of blood testing based on the Rhesus, Kell and Duffy blood-grouping systems. This test eliminated only 40 per cent of the entire male population at a time from being the possible father, hence making it slightly more accurate.

It wasn't until the 1970s that it became possible to exclude a significantly higher percentage of the male population through paternity testing. This new method, known as HLA (Human Leukocyte Antigens) testing, eliminates 80 per cent of the male population from being the possible father. In some instances, it has been possible to produce a probability of paternity of up to 90 per cent. However, HLA testing cannot differentiate between alleged fathers who are related.

In the 1980s, the technology behind DNA testing was ramped up further with the introduction of RFLP (Restriction Fragment-Length Polymorphism) testing, whereby DNA is extracted from blood samples by a process of purification that reduces the sample down to a clump of long, string-like molecules. This test, while being conclusive, is an old technique that requires large samples and a long processing time.

The 1990s saw the introduction of the testing method we use today, that of extracting DNA via PCR (Polymerise Chain Reaction). With this method, scientists can extract millions of copies of the subject's DNA – each taking the form of a long, string-like molecule – from a single small sample, such as a cheek swab. The PCR method creates copies of only a small fragment from this molecule, and scientists can copy the exact part of the DNA molecule that they require. Once obtained, the DNA fragment can then be analysed. The accuracy of this test has an exclusion rate of over 99.99%.

THE SCIENCE OF DNA TESTING

DNA is a group of chemicals in which is encoded the fundamental genetic blueprint that determines a person's biological characteristics. It's present in every cell of the human body, and indeed in the cells of most other forms of life on

4

this planet. Because of this, and primarily because of new technology, the samples from which it can be extracted aren't restricted to blood samples. The most common method of obtaining DNA samples is by using 'buccal swabs' (shaped like a cotton bud), which are rubbed gently across the surface of the inner cheek to gather up loose skin cells. Some testing centres advertise the fact that they can extract DNA from anything from hair follicles to used handkerchiefs, but the buccal swab is the more accepted method and also makes the extraction process easier.

Each child inherits 50 per cent of their DNA from their mother and 50 per cent from their father, and so, in order to carry out an accurate test, samples are required from the mother, the father and the child. (From a technical standpoint, the mother's samples are not required but are nevertheless recommended. Besides, the UK paternity-testing code of practice recommends that DNA testing companies get consent from both parents for testing purposes.) The DNA of the child is first compared with that of the mother and the common parts are then identified. The logical

5

assumption then is that any remaining DNA in the child's profile can have originated only with the biological father. The DNA of the supposed father is then analysed and compared similarly.

In simple terms, the test will show whether or not the child and supposed father share common genetic material. In other words, the supposed father will be either included or excluded by the results. The test can establish whether the presumed father is the biological parent, a relative or completely unrelated. It's a very accurate method of testing, the only possible confusion arising when the father is one of a number of identical twins, triplets, etc.

It's essential that as much information about the particular reasons for doing the test include the relationships – suspect or factual – of all the participants in the testing process, as there have been recent cases where companies offering a paternity-testing service have fraudulently guessed the results in the absence of any legitimate scientific testing data. At the time of publication, there has been only one such occurrence in the UK that has been identified and made public. In the field of DNA testing, as with all other industries, while the majority of companies are honest and professional, there are of course unscrupulous individuals who work in an unethical, unprofessional manner. It's therefore advisable that anyone looking to obtain a DNA-based test should research the companies they're looking to approach, asking questions such as:

How long has the company been established?
Is the company allowed to provide tests for the court?
Do they follow the government's code of practice?

Most legitimate companies use professionally recognised laboratories, as these provide safeguards against the switching or contamination of samples being examined. In short, once you've received your results from one of these labs, you can be very certain that they illustrate accurately the parentage of the child involved.

WHY TAKE A PATERNITY TEST?

Paternity testing isn't just about finding out the identity of the father of a child. Individuals of all ages have approached companies providing DNA testing, and for a wide range of reasons. One American woman tells a story that's not uncommon:

'I need to let everyone know that having a paternity test saved my life! I was diagnosed with cancer, and thought that my life was over. Due to my particular form of cancer, I required a special blood transfusion. Having no mother and not being sure who my father was, I thought that I might die. I was relieved when one of my friends told me of a company who could do a DNA test that would help me answer my question and, perhaps, save my life. One of the men who we thought was my dad agreed to be tested, and we found out that he was my biological father. I had the transfusion and now I am fine, which means that not only do my children have a mother, but a real granddaddy too.'

This real-life scenario demonstrates that establishing a biological relationship had an extremely positive impact on this person's life – and, in fact, probably saved it!

Other individuals have used the DNA-based testing services to establish immigration rights and settle marital disputes. A well-known commentator on the subject, Barry Pearson, has identified one of the most common motivators of DNA testing to be the search for 'peace of mind'. In these instances, he acknowledges, 'Some men want to know if the children they are told are theirs really are their children. The normal method available to find out is a DNA paternity test. Some of these tests can work with just samples from the man and the child.'

However, as English law stands at present, tests carried out without the consent of a mother or a relevant court order are not recognised in court. In later chapters in this book, I will explain how an individual should approach DNA-based paternity testing and the possible ramifications of the results.

2

DIFFERENT TESTS AND HOW THEY ARE PERFORMED

There are currently a number of DNA-based tests that can be carried out in order to establish biological relationships. The type of test employed will depend on each individual's situation. And, as I mentioned in the previous chapter, while the media will often portray paternity testing as a tool used by irresponsible men hoping to avoid the financial burden of child support, the truth can in fact be far more complex.

Recent high-profile cases have shown that the need for such testing exists in the legal arena. In one case (*Hurley v. Bing*), it enabled a mother to force her child's father to pay maintenance contributions, while another highlighted the situation where a father wishes to prove the paternity of a child in order to establish a right of access (ie *Blunkett v. Quinn*). A final example is the case of *Snowden v. Fry*, where a child has been able finally to put questions about their parentage to rest.

The majority of cases are based on the peace-of-mind idea, in as much as doubts about a child's heritage – whether in the mind of the child him- or herself or of either one of the parents – have caused issues in the past that can at last be simply and quickly resolved. It is not, however, only the parents who can be tested; cousins, siblings and grandparents can also offer proof of an individual's genetic identity. Figure 1 shows how the participants in a DNA-based paternity test can be related to the child.

Fig. 1 – Possible relationships in a paternity test

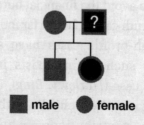

■ male ● female

THE PATERNITY TEST

Perhaps the most common and best-known situation in which a DNA test is sought involves checking a suspected father's DNA against that of a child. As mentioned in the previous chapter, this is done painlessly by obtaining a sample using the buccal swab. The results of such tests are reported as a percentage of likelihood, and will almost always be either 0 per cent or greater than 99 per cent, the former effectively excluding the alleged father from any putative biological relationship and the latter proving that the relationship does exist. Such results are often relied upon in child-custody disputes, maintenance disputes, immigration matters and birth-certificate changes. If the individual does intend to rely on the test result in court, the samples will generally have to be obtained in the presence of a GP or other professional, such as a nurse or midwife, so that, if there is a further dispute about the reliability of the samples, a complete trail can be established in relation to where the samples originated.

While the results of many such tests are used in court proceedings, there are a vast number of individuals who are simply looking for peace of mind (also referred to as DNA tests for 'personal reasons), as demonstrated in the following account:

'I have often thought about how different things might be now if we had not gone through with the DNA testing. Instead of *knowing* (with better than 99.9% accuracy) that I *am* the father of the most beautiful little girl in the world, I would probably still be trying to believe I am her father and regretting that my parents showed no interest in a child that was of dubious genealogy. (Heaven knows the challenges of parenting are enough without having to endure a sea of doubt!)

'In our case, my partner told me that she was certain from the beginning of pregnancy that the baby was *not* mine, even though the margin of error in her calculations seemed far from certain to me.

'As time went on, the "suspected" father mistreated my partner and she became convinced that – genetically or not – I was a much better partner and father for her child. I had not planned these circum-

14

stances but was willing to try to do my best at being a father, in any case, and take it all one day at a time. We initially discussed testing, but my partner was afraid for [the following] reasons: One: if tests proved that her ex-boyfriend was the father, he might try to steal the child or his partner might even try to harm the child. Two: if tests proved I was not the father, I might lose interest and leave.

'The more I thought about it, the more I became convinced that knowing the truth is always better than living in doubt. Furthermore, why should I deny myself, my parents and the child the possible joy of knowing I was the father?

'I knew there was no way to deny that I would feel differently toward the child if I knew she was "mine", no matter how much I loved her otherwise. And if I really was the father, what a terrible and silly waste it would be to live willingly in ignorance of this!

'I knew it was important for everyone to know the truth, but my partner needed a little time to get over the fear of taking that step into the unknown (even though we were already living in it!). We discussed taking a blood test, but didn't want to subject the baby to such a painful experience. We had no idea that DNA testing was such an affordable and painless option until a friend suggested it.

'The results the test yielded brought us more joy than we ever imagined. Not only does my daughter now have the adoration of her grandparents, but also this has given me a better relationship with my mother than I have had in many years.

'I will never forget the day I got the results back. I relive the joy every time I look into my little girl's eyes!'

PRE-NATAL TESTING

This method of DNA testing is used in a number of circumstances, such as:

Rape cases – ie instances where victims of sexual assault want to be certain that the pregnancy was not a result of the attack but rather from a partner. The test will allow for an informed choice to be made by the victim in relation to the pregnancy.

Affairs – ie the participant has had an affair and needs to be sure that her husband is the child's biological father.

In some circumstances with regard to the latter scenario, the husband might accuse his spouse of an affair, and the test can be used to argue against this.

There are two methods employed to check the paternity of a child while it's still in the womb. The first is known as CVS (Chorionic Villus Sampling) and is performed in the very early stages of pregnancy, usually between the ninth and thirteenth weeks. It involves inserting a catheter through the cervix and removing a small sample of cells from the gestation sac by gentle suction. It is reported that, after undergoing this procedure, some patients experience discomfort, but the results obtained via this method are just as accurate as those obtained from tests carried out on a child after he or she has been born. With this method of testing, the DNA is extracted from the CVS samples, after which the process is the same as that used to extract DNA from a cheek swab during regular testing. Once the DNA is extracted, the testing process continues as normal.

The second method requires something known as an *amniocentesis test*, performed in the later stages of pregnancy. This procedure involves withdrawing a small amount of amniotic fluid (ie the fluid in which a foetus is immersed) through the abdomen with a needle and syringe. Like the extraction of DNA from the CVS samples, the DNA extracted from the amniotic fluid is then used to process the DNA test.

Both tests provide results that are conclusive, as long as the DNA-extraction procedure is successful, and both should be carried out by trained medical professionals, as there is some risk to the foetus in each case. Studies have indicated that the chances of losing a baby increase by 1% with the CVS method and 0.5% with amniocentesis. Obviously, anyone considering such tests should seek advice from a doctor prior to undergoing such tests.

Many people may wonder why an individual would put a child at risk with pre-natal testing, and in this respect the following story reveals why one woman felt that it was so important to make that difficult decision to ensure the paternity of her child:

'There is really no easy way to say this, just that simply, one night, as I was working late, I was raped by a colleague. I didn't tell anyone at the time and, though it may sound daft, the thought of getting pregnant didn't cross my mind. It wasn't until I missed my period that the fear set in. My husband is a very kind and loving man, and I am sure that if I had told him from the beginning then he would have helped me deal with the crisis I was faced with, but as time went by I found it harder and harder to say anything. I knew that I needed to know if this was a child made within a loving relationship or whether I ran the risk of giving birth to a rapist's baby.

'The doctor was very supportive and guided me through the CVS process. Afterwards I had a bit of bleeding and some tummy ache, but that was all. The results were back in three days, and thankfully our little girl is my husband's daughter. Later on, I found the courage to tell my husband what had happened and have had counselling to help me come to terms with everything. I am just so glad that I could have the test before the birth. The thought of having to go through labour and give birth to that person's baby after what he had already put me through was enough to make me think about suicide.'

MATERNITY TESTING

Fig. 2 – Possible relationships in a maternity test

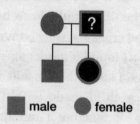

■ male ● female

As demonstrated in Figure 2, the participants used in a DNA-based maternity test are the same as those involved with the paternity tests. And, while maternity testing is rare, there are some cases where it is appropriate – for example, to confirm that a child previously given up for adoption has found his or her correct birth mother or, in even rarer cases, confirming that a hospital hasn't mistakenly mixed up the mother's baby with someone else's.

In some cases where people have undergone IVF treatment, the prospective parents might wish to ensure that the correct egg has been used in the process of conception. This testing can be done either before or after the child is born.

SIBLING TESTING

Fig. 3 – Possible relationships in a sibling test

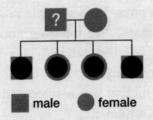

■ male ● female

Sibling tests can establish the probability of there being a biological relationship between two people, commonly when there are no other relatives available for testing. Figure 3 illustrates the participants that might be involved in such a process. The tests can establish full- and half-sibling relationships, possibly putting an individual's mind to rest after years of doubt about their parentage. People of all ages are now turning to DNA testing to find out if stories concerning possible family members are really true. One woman waited almost thirty years to find out who her family really were.

Sibling testing is commonly used to determine whether two individuals share common parents, as demonstrated by the following account:

'I'm a member of a large family – four girls and three boys. When people met us they would remark, "It's amazing none of you kids look anything alike." This has been debated by many over the years, and when asked my mother would often remark, "Oh, it was the milkman."

'In my teens, my siblings and I learned that my oldest brother was really our half-brother, the product of a previous marriage my mother had had before the Second World War. Likewise, we had always assumed that my older sister (I'm number three) was my father's child, as we had always heard stories about them being sweethearts before the war and then marrying afterwards. (We were kids; we really didn't think too much about the

22

actual timeline for all of this.) When my older sister went to get her marriage licence, she found that someone other than my dad was listed as her father. Even after that, it was still rumoured that my father was indeed her father, but because of the times my mother found it necessary to list her now ex-husband as the father.

'For years, as I looked at my siblings, it was obvious that I took after my mother's side of the family and not a trace of my father's side could be seen. My fair complexion and blue eyes greatly contrasted with the darker eyes and colouring of my siblings. As I got older, I learned that my mother had lived on her own near her parents before marrying my father. Much to my surprise, I discovered this while we were going through all the files after my father's death. My mother had died eight years earlier, when I was twenty-three. I started to question whether or not my dad was really my biological father or not. This doubt persisted for years, so I finally asked my mother's sister if this might be the case. She indicated that she knew of nothing like that, but that question had put the wheels in motion.

'Later that year, at Christmas, I received the usual Christmas card from one of my other aunts, only this time it contained a short letter. In it she told me that my "real" father's name was Lars, or Larson, and that he had been of Swedish decent. I was stunned, but it seemed to explain so many things: my colouring and even the spelling of my first name. I called my sisters with the news; they were stunned as well. My older sister said, "What does it matter? Daddy is Daddy. Besides, what can you do about it now? They're both dead; you can't ask them." When I mentioned the rumours about her paternity, she indicated that she had no desire to question it at all.

'I, on the other hand, continued to be curious. I tried to corroborate this bombshell from my aunt, but it seemed that she was the only one who had any first-hand knowledge about the matter. Supposedly, my mother shared her secret with her sister before I was born and after

24

she had married my father. The others had only hearsay from my aunt on this.

'I wouldn't say I became obsessed with this but, unlike my older sister, I did want to know, one way or another. Since both my parents were dead, I didn't think DNA testing would be possible. I finally decided to check things out to see if one could determine paternity using a sibling's DNA. I thought it might be possible to determine the likelihood, but I didn't think the probability could be conclusive. Much to my surprise and delight, I found that this was indeed possible, and to a very high degree of probability. So I requested a DNA testing kit and submitted samples from my younger sister and me (my older sister wanted nothing to do with it). Imagine my surprise when the results came back indicating a 99.99% probability that we shared the same two parents. Now the only mystery is either why my aunt made up this story or why my mother told her sister such a story.

'It's sad to think that, if my mother *had* believed the story to be true all those years, she might also have lived with that guilt that I was another man's child and thought that she'd deceived my father. My aunt said that she didn't think my father ever knew or suspected what my mother had thought was the truth, but then again my aunt would have had no way of really knowing.

'So now the big family secret turns out not to be true. Without the DNA testing, I would have been left wondering for the rest of my life. The only question that remains is the original one I had to begin with: why am I so different from my siblings?'

GRANDPARENT TESTING

Fig. 4 – Single-grandparent test

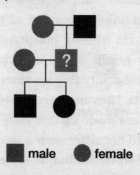

male female

There are two methods commonly used for testing the probability of whether someone is or isn't a grandparent of a certain child. Figure 4 illustrates one of these, 'single grandparent' DNA analysis, which involves the participation of one grandparent and one child. The presence of a parent is optional but strongly recommended. The results of this test may vary between 1 per cent exclusion and 99 per cent inclusion, but they generally come within the range of less than 15 per cent and greater than 90 per cent exclusion and inclusion, respectively.

Fig. 5 – Duo grandparent test

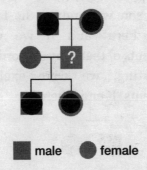

■ male　● female

The test illustrated in Figure 5 requires the participation of both grandparents along with that of the child and, preferably, one parent.

Where both potential grandparents are available, the results are highly accurate, giving a reading of greater than 99 per cent inclusion or 0 per cent exclusion.

This kind of test is useful in cases where grandparents wish to obtain access rights to a child but the father is absent and they are on less than amicable terms with the mother. A grandparent could also be the only available possible relative in an adoption situation. The following story shows how effective the results are, and perhaps how unwelcome:

'We simply wanted to know if our deceased son was indeed the father of a baby in Peru, and if we were the grandparents of that baby. According to the DNA testing, we were excluded as grandparents. It ends there.'

EXTENDED FAMILY TESTING

Fig. 6 – 'Aunt/uncle' DNA test

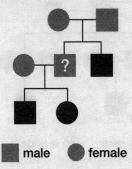

■ male ● female

Figure 6 illustrates the 'aunt/uncle' DNA test (conducted via a process known as 'avuncular testing'), which determines the probability of two people being relatives. In particular, it measures the probability of Participant 1 being either aunt or uncle to Participant 2. The conclusive range here is less than 20 per cent exclusion or more than 80 per cent inclusion, and the probability increases greatly if other known relatives, such as the parent of the niece or nephew, are available for testing. Results are then enhanced to between 1 per cent exclusion and 99 per cent inclusion.

Fig. 7 – First-cousin analysis

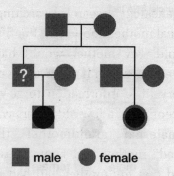

male   female

Figure 7 illustrates a testing method known as the 'first-cousin analysis', designed to measure the probability of two people alleged to be first cousins actually being related. The accuracy of this test can be narrowed down to less than 20 per cent exclusion or greater than 80 per cent inclusion.

While it might seem unlikely that neither parents nor grandparents are available to participate in testing, there are times when doubts about someone's genetic heritage have persisted for several decades before answers are sought. The following story illustrates how external-family analysis helped one client to find out about his genetic history:

'A little over a year ago, I started doing some genealogy research, searching for my biological mother and father. I'm fifty-seven years old. My mother would have been almost eighty-seven if she was alive today. I had very little information to go on, with the exception of a few old letters written by my maternal grandmother that my "adopted" mother had kept. These letters inferred that my mother had had relations with a man who was in the military during World War Two, during which time she got pregnant and gave me up for adoption. She was more than likely not married.

'From some of the information contained in these letters, I was able to locate individuals with the same family name still living in the town from where they originated. After determining that one individual might be a cousin, I asked if he would agree to do a DNA test with me. When he accepted, I searched the web and found a test kit, which was sent to the potential cousin and myself. The results from this test, the DNA haplotype analysis, indicated that a common maternal-line inheritance was supported. This was astonishing news, not only to myself but also to the group of newfound cousins.

'There were two siblings of my mother left alive in this town. One of them, my aunt, told her children that her sister had never had any children and asked them to tell me to leave them alone. The other, an uncle, when confronted with the information by his daughter, acknowledged that his sister had had a child. Two days after he made this admission, he and his wife were both killed in an auto accident.

'I couldn't have done a test like this a few years ago, so the timing of my search couldn't have been better. I still don't know who my father was, but I'm determined to find more information.'

③
HOW ACCURATE ARE THE TESTS?

Laboratories used by recommended DNA-testing companies follow strict guidelines in the storing and testing of the samples they receive. Most will carry out at least two tests on the samples provided in order to ensure that the results are accurate. In many cases – particularly peace-of-mind or personal tests – the collection of samples, upon which accurate readings are dependent, is the responsibility of the individual commissioning the test. Essentially, the testing process will take place on the samples that are received by the laboratory, and there is no chain of events and witnessing of sample collection that can be relied upon if the results are challenged.

SECURING THE CORRECT SAMPLES

In cases where a court has ordered a test to be performed, a trained professional will collect the necessary samples. Otherwise, many companies now provide 'home kits', which are simple and non-intrusive methods of obtaining samples. Each kit generally contains a set of sealable cheek (ie buccal) swabs, an instruction leaflet detailing how to collect the samples and from whom, and a consent form that needs to be signed in order to authorise the laboratory to test the samples.

Testing can be carried out on hair and other bodily substances, but the cheek-swabbing process is far less complicated for the untrained, and, with rising public awareness that cases can be solved without resorting to court proceedings, home testing has become much more common over the last decade. Individuals have, in general, found home testing kits to be simple and easy to use, as described in the following account:

'We had the test run to confirm that my stepson was indeed my husband's child from his first marriage, as she would not or could not confirm it for us. We got the test kit and within days did the test. After we collected the samples, we waited patiently for the results to be released. When they finally came, we were overjoyed to find that our son is indeed my husband's child. It was such a relief to find out without having to deal with his ex-wife. We are completely convinced that the testing process was the right thing for us to do as a family and so grateful that the simple home kit helped us to find the answer to a long-overdue question – eight and a half years overdue, to be exact! I have told everyone all about it, and they can't believe how quick and easy the whole process truly is. It's wonderful to at last have peace of mind.'

ENSURING NO SUBSTITUTION

Obviously, in cases where an individual is testing themselves and their presumed child privately in order to obtain results for their own peace of mind, there shouldn't be any sample substitution, as that would defeat the object of the exercise. However, in cases of disputed paternity, there have been cases in the past that have resulted in legal proceedings against individuals who have obtained samples from a third party or have sent an unrelated person to the test centre in order to ensure that the results would not confirm the true parentage of the child in question (for instance, where a father is attempting to avoid being responsible for providing financial support). While these occurrences were rare, the courts have now put into place systems that make substitution of samples almost impossible.

In disputed cases, particularly those in which the CSA (Child Support Agency) is involved, only after the identities of all presumed and actual parents have been verified will a doctor gather the necessary samples. Once these have been obtained, they are then subject to a strict line of custody and are sent directly to the laboratory for testing. At no stage are the samples accessible by the public.

The identities of those undertaking the testing process are usually established by reference to passports or other photographic identification, while sometimes, in order to eliminate all possible complications, testing companies will also require fingerprints as a form of identification. The samples are placed in a sealed, tamper-proof envelope, signed by the collector, ensuring that, if the result is challenged, the court can review the testing process, from collection of samples to results. Sometimes when results are disputed, the court might ask for further tests to be performed by an independent company.

Of course, disputed paternity cases don't have to become unpleasant, as the following testimony suggests:

'I am twenty-one and my son's father is thirty-five. When I got pregnant, he didn't want to believe that my son was his, and I didn't want to take him to court for fear of creating bad feeling that could possibly last forever. I thought about it for a long time. Finally, when my son was six months old, I really wanted him to know who his father was, so I started researching how we could have a private test done. That's when I came across a company on the internet that provided home kits. I was really excited because it meant we could both know and not make it a legal issue. In the end, it turned out that he was the father, and now we live together, as friends, with our son.'

POSITIVE/NEGATIVE – HOW SURE ARE THEY?

DNA testing is, on the whole, conclusive and, in cases where there is a straight paternity issue, the test will show whether the male involved is the father (100 per cent accuracy) or isn't the father (over 99 per cent accuracy). There is no middle ground, as the child's DNA can be inherited from only its biological parents.

CHOOSING A REPUTABLE TESTING COMPANY

In 2003, one highly publicised case illustrated how easy it was for an individual to set up a company advertising DNA-based testing services while not actually having any means of testing samples. While it's important to note that the majority of testing companies are reputable and can be trusted to provide accurate results from samples submitted to them via posted home testing kits, it's important to establish whether or not the company you're choosing is legitimate and reliable.

Simple methods of ensuring that the company of your choice is trustworthy include:

• Checking that the company uses ISO (International Standard Organisation)-accredited laboratories. Don't be afraid to ask which laboratories the company uses and where these are situated. An ISO accreditation indicates that the laboratory operates to the highest European standards of quality with very strict quality control.

• When browsing through the website, look for the organisation's company registration number. Also, ask them if they have a VAT registration number; if they do, this implies that the company has sales of £52,000 or more per annum, indicating that they are well established. DNA testing can change the life of whoever undertakes it, as well as their family, so it's essential that you feel comfortable with the company you select.

• Do you have the option to visit the offices, either to deposit a sample or to ask further questions?

- Do you have the option to visit the laboratories?

- Are they affiliated to other organisations offering sources of information and support? This information should be present on their website.

- Are they happy to answer any and all of your questions?

- Do they refer to the British government's voluntary code of practice?

- Are they recommended by the DCA (Department for Constitutional Affairs)?

- An up-to-date list of companies recommended by the DCA is obtainable from the department itself.

4

THE REPORT

If you're in the process of having your DNA tested, after the laboratory has tested your samples, they'll send you a report similar to the one below:

Relationship	Biological Mother	Child	Alleged Father
Name:	A mother	A child	A man
Date of Birth: 07/11/1967		8/12/02	05/12/69
Ethnicity		White	
Collection Date:			
Results: Mr A Man is not excluded from being the biological father of A Child			

Summary and interpretation of results: Based on the results from the sixteen genetic systems (loci) shown below, Mr A Man is not excluded from being the biological father of A Child because they share alleles for all of the fifteen loci analysed. The Combined Parentage Index value of 11,424 is the product of the Parentage Index values and indicates that Mr A Man is 11,424 times more likely to be the biological father of A Child than an untested, unrelated person of the same ethnicity. Assuming a prior probability of 50%, the probability of paternity is 99.9912%.

Locus	Biological Mother Not Provided	Child A Child	Alleged Father A Man	Parentage Index
AMEL	x	X	X, Y	-
CSF1PO	-12	10, 12	10	2.065
D2S1338	-17,22	17, 23	23, 26	2.182
D3S1358	16,18	15, 17	17, 18	1.374
D5S818	-12,13	12, 13	12, 13	2.326
D7S820	-10,11	10	10	3.674
D8S1179	-11.15	14, 15	11, 14	1.171
D13S317	-8,11	8, 13	12, 13	2.238
D16S539	-11,13	12, 13	11, 12	0.827
D18S51	-15.17	14, 17	12, 14	1.492
D19S433	-13,14	13, 15	14, 15	1.586
D21S11	-28,30	30	30	3.967
FGA	-22,24	22	20, 22	2.957
TH01	-8,9.3	7, 9.3	6, 7	1.148
TPOX	-8,10	8, 11	8, 11	1.433
vWA	-16,19	16, 18	16, 18	2.182

This information is confidential. Deoxyribonucleic acid (DNA) was isolated from the specimens submitted for each of the study participants and was characterised using the polymerise chain reaction (PCR) for the genetic systems (loci) listed above. The associations of name, relationship and ethnicity with the accompanying results rely strictly on the information provided to the laboratory. The company is not responsible for the origin and/or transportation of the samples prior to arriving at the laboratory. The collection of specimens for the purpose of generating the data shown above was not performed in compliance with established chain-of-custody guidelines. Therefore, these results might not be admissible in a court of law. The participants of this study understand and agree that these results are strictly for personal information only.

Once you have received your report, you should feel free to contact the company that ordered the test if you need to have it explained in further detail. Almost all reputable companies will take the time to explain your report and the meaning of its content, as well as how the results will affect you.

In cases where the court has commissioned a test for disputed paternity, the report will be written in a more formal manner, readings something like this:

'Combined paternity index: 403,257

Probability of paternity: 99.99%

The putative father, A Man, is not excluded from the paternity of A Child. The results are consistent with A Man being the biological father of A Child. Therefore, the probability of paternity, given the DNA evidence, is at least 99.9999%, compared to an untested random male resident in the UK.'

The wording of this kind of report is applied carefully in order to comply with the Family Law Reform Act of 1969 and thus be admissible in court.

5

SOME FREQUENTLY ASKED QUESTIONS

Can you test a child who is less than six months old?

Yes, you can. DNA is present from conception, and viable samples can be taken from unborn children as described in Chapter 2, 'Different Tests and How They Are Performed'. In the case of newborns, blood can be drawn from the umbilical cord, causing no discomfort to the foetus. Older babies and young children can be tested using a cheek swab.

What if the person being tested has had a blood transfusion?

This will make no difference because, in the testing process, DNA is extracted from only white cells, while transfusions involve the transfer of only red blood cells from a donor to a recipient. Therefore, blood samples collected from a transfusion patient won't contain the DNA of the donor. However, tested parties will have to indicate their transfusion history in the rare event that a court orders a red-blood-cell test.

Because white blood cells are made in the bone marrow, however, blood collected from individuals who have had a bone-marrow transplant might produce DNA results that point to the identity of the donor rather than that of the recipient. Therefore, such transplant recipients should be tested via specimens other than blood, such as cheek epithelial cells, taken with buccal swabs.

Do drugs affect the test?
No. Alcohol, recreational drugs or any other form of medication won't affect the test.

What kinds of samples are required?
In most cases, cells from the inside of the cheek are used, which produce results just as accurate as those from tests carried out on blood samples, as all cells in the body contain DNA.

How many tests can I have?

As many as you like, as there's no way you're ever going to run out of DNA! Families sometimes prefer to perform two independent tests to give them additional peace of mind.

Can you identify genetic disorders or other illness from such DNA testing?

No. If DNA is collected for the purpose of establishing a biological relationship, the company doing the testing will usually look at only those genes that indicate the subject's paternity. Some companies might perform an additional an analysis for genetic disorders on request, but for this you'd more than likely need to be referred by your doctor. Some web-based companies offer genetic-disorder testing and gene screening, but this is a very sensitive area. For instance, imagine you were told that you had a 35 per cent chance of contracting breast cancer. How could you make an informed choice based on this type of percentage?

How fresh do the samples need to be?

Archaeologists have extracted viable DNA for testing purposes from samples hundreds of years old, so there's no real concern here. Where DNA is being tested for immigration purposes, for example, sometimes samples will be collected in remote countries and will take a while to reach the lab.

What if the potential fathers being tested are related?

If this is the case, the men being tested will share some of the same genetic markers, making it more difficult to distinguish between their genetic profiles than between those of unrelated men. For example, a man shares 50 per cent of his inherited genetic markers with his brothers, sons and father; 25 per cent with his uncles, nephews, half-brothers and grandfather; and 12.5 per cent with his first cousins. If two identical twins are being tested, however, genetic tests cannot distinguish between them.

In the standard statistical analysis of test results, it is assumed that the possible fathers are not related. If there is any reason to believe they are related, this should be brought to the testing company's attention from the outset so that appropriate testing and statistical analysis can be performed.

48

Who do I contact to get a DNA-based paternity test?

If you wish to be tested privately, your GP should be able to refer you to an appropriate testing company, while the Department for Constitutional Affairs will be able to recommend an appropriate private company.

In March 2001, the British government issued a code of practice to be adhered to by all paternity-testing services. Only those organisations that comply with this code of practice are on the list of approved testers empowered to carry out court-ordered tests.

How much does a test cost?

Details of approved testers can be obtained by the court ordering the test, although testing prices vary among the private companies that perform them, ranging from £120 to around £350, depending on the type of test you require. One should also be prepared to pay for collection fees charged by the GP or Nurse. The Citizens' Advice Bureau state that there is no maximum fee for DNA testing, but realistically one should expect to pay around £150 per person to be tested, and of course it's a good idea to find out how much each test will cost before having it done.

How long does it take to get the results?

On average, the results of each test will be returned after between seven and ten days. However, for an additional charge, many companies can provide results more quickly than this.

6

EMOTIONAL AND LEGAL IMPLICATIONS

Deciding to take a paternity test is only the first step. Before you proceed, you should consider carefully your reasons for taking one and the consequences the results will have on your life and all other people likely to be affected by them – children, parents, grandparents, siblings, etc.

As I mentioned earlier, paternity tests aren't taken only by men looking to establish whether or not they've fathered a child, and, in cases where an adult wants a 'peace-of-mind' test or wants one performed for medical purposes, as described earlier, the possible emotional fallout is perhaps not as damaging as from those tests taken for other purposes. This is because, in most cases, there has been a lifetime of suspicion about biological relationships, as illustrated perfectly by the following case:

'My husband is thirty-four years old. He has always known who his father was, but for some reason I never understood his father has never had any kind of relationship with him. However, after thirty-four years it all makes sense. Someone told my husband that the man who he thought was his father really wasn't. He confronted his mother about it and, after denying it for a couple of days, she finally admitted that there had been an affair. Upon hearing this, I contacted one of the men and asked him if he'd had a relationship with my husband's mother in the past, and he confirmed that he had. I told him that a child had been conceived thirty-four years earlier from a relationship my mother-in-law had had and that I thought he might be the father. I asked him if he'd take a DNA test with my husband and mother-in-law, and he agreed. When the results came back, it turned out that there was a greater than 99.8% chance that this man was my husband's biological father.

'Since finding out this result, almost a year ago, my husband has had a relationship with his true biological father, which has given him closure on an open wound that had troubled him for some time.'

In cases like this one, fraught with unanswered questions and legitimate feelings of rejection, the identification of an actual biological father can be only beneficial.

Other, more procedural reasons for taking a paternity test, such as for immigration purposes, are relatively straightforward in comparison, as the following case shows:

'A friend of mine was married in China, and he and his wife had a boy a few months later. They wished to immigrate to Hong Kong because he's a citizen there, but it was difficult to prove that the newborn boy was indeed his. The paternity DNA testing helped him to prove that the child has the same DNA profile as his father. This satisfied the immigration requirement and they were able to get residency status for their son.'

However, even when tests are carried out for immigration purposes, they can occasionally give unexpected results. There have been instances when, during a routine immigration process, fathers have found out that a child isn't actually theirs, in once case some sixteen years after the child's birth.

In cases involving young children and presumptive fathers, there is often a great deal of dispute. The British Medical Association's Ethics Department published the following guidelines relating to 'motherless testing':

'The Human Genetics Commission points out that motherless testing could have serious consequences for family life if large numbers of men decided to check whether or not the child they are supporting is genetically theirs. Where the putative father has parental responsibility for the child, such testing could be undertaken without the knowledge of the mother. The BMA believes that this could be very harmful to the child, as well as to the family unit as a whole, and would prefer to see a situation in which the consent of the mother and putative father (and the child, if sufficiently mature) is required for paternity testing. In the absence of such a requirement, where doctors are consulted they should encourage those seeking testing to discuss their plans with the child's mother, and the BMA would advise doctors not to become involved if that advice is rejected.'

As I indicated earlier, motherless testing is not recognised in court. However, with access to accurate home testing kits, so-called 'secret testing' has become more common. At present, a presumptive father can collect DNA from his child and send the samples to a laboratory for testing, and, if the results show that he isn't the actual father, the CSA can be petitioned to carry out a second DNA test to establish the child's true paternity. As Barry Pearson – a well-known commentator on this subject – argues, 'These unofficial motherless paternity tests will tend to strengthen the vast majority of families where they are used, giving those families a better prognosis. That justifies the alternative name for such paternity tests: peace-of-mind tests. The minority of families where such tests may prove disruptive are not marriages made in heaven, and it often isn't obvious whether the children would be best off if the family remains intact or if it separates.'

What follows is a series of questions you should ask yourself if you're thinking of taking a paternity test.

Why are you taking the test?

The following testimonies demonstrate two very different reactions to individual paternity tests.

'I recently found out that my wife had been with another man a few times in the early years of our marriage. I wasn't exactly a saint back then, either, but all of this history was a long, long time ago and we truly forgave and forgot. However, our son had been conceived and born around that time, and I kept wondering whether or not he was my biological son. I tried to forget about it, thinking that I had raised and loved him, so it didn't matter, but the idea kept eating away at my mind and was impacting on our relationship, so I decided that the only way to move forward was to get closure by doing a DNA test.

'It's impossible to imagine the relief I felt when I read the report, which stated that I am indeed his father. My nagging doubts are gone for good, and we couldn't be happier. Finding out allowed me to not hold back with my emotional bond, which I had been doing for years without ever even realising it.'

'I had a genetic test a year and a half ago and found out that my daughter wasn't even mine. Now I have no access and no daughter. I didn't consider the consequences of the test and, had I known that I wouldn't see her again afterwards, I might have reconsidered going through with it.'

In Australia recently, a suspicious mother-in-law, convinced that her son wasn't the father, had her son's youngest child tested secretly using a home kit. She was right. Moreover, on testing his three remaining children, they found that her son was the biological father of only one child. He separated from his wife and initiated a case for 'paternity fraud'. However, regardless of whether or not he was the children's biological father, he had invested emotionally in each of the children for almost fifteen years. As far as the children were concerned, he was their father and their mother's mysterious lover was nothing.

In his paper 'The Natural Father: Genetic Paternity Testing, Marriage and Fatherhood', Gregory Kaebnick suggests that parenthood is a psychological relationship rather than a biological one, and that testing should perhaps be limited to that ordered by a court. Barry Pearson, however, disagrees: 'Unofficial paternity tests deliver knowledge about oneself. If the law bans them, this will be a rare example... where seeking the truth about oneself is a crime.'

In cases where a man has assumed that he is the father of a child, however, and they have a loving relationship, a test that eliminates that man as the father of the child will undoubtedly result in the nature of the relationship changing, even if the man who believed himself to be the father is the only person with access to the result.

What do you hope to gain?

Contact? Release from financial commitments? The truth? Examine closely your reasons for taking the test. Speak to a therapist, contact organisations such as Families Need Fathers or Parentline and discuss the ramifications of the possible results with the child's grandparents. And, of course, think about how the test results could affect your relationship with the child.

Unless it has been ordered to be carried out by a court of law, it's ultimately your decision as to whether or not you need to take a test, and you'll have to live with the consequences.

Who needs to know the result?

With those tests not taken under a court order, the result is for the commissioning party only. If, for instance, an adult has somehow taken a sample from a suspected father or other relative after the father or other relative has refused to undergo such a test, the legal implications are immeasurable.

However, in the case of a secret paternity test, where an individual sharing custody of a child performs a motherless test and is excluded from paternity, official bodies such as the Child Support Agency can be made aware of the situation so that they can order official tests for their own purposes.

Can the test results be used in court?

The results of tests carried out with home testing kits aren't admissible as evidence in court. Only those samples gathered by a doctor or other recognised professional can be used to establish paternity in court.

I'm not the child's legal guardian. How do I get him/her tested?

In these cases, tests can be procured only through legal proceedings, either through family courts or, if paternity is being disputed while paying child support, by petitioning the CSA. In the latter case, it's inappropriate to perform a secret test using a domestic testing kit.

Can my child be forced to take a paternity test?

If you've secured a court order then, yes, the child will have to take a test. However, this will generally be carried out via a non-intrusive cheek swab – harmless and completed in a matter of moments.

Can my ex-partner be forced to take the test?

Again, only once you've obtain a court order to prove or disprove paternity. In the UK, if you refuse to take a test, the CSA will assume that you are the child's parent and expect you to contribute financial child support.

Will my DNA be made available to other official bodies, such as the police?

No. Under the UK Data Protection Act, access to both samples and results is strictly limited. The companies and laboratories involved in DNA-based paternity testing are legally required to keep all information pertaining to their clients confidential.

THE CSA AND CARCASS

7

THE CSA AND CAFCASS

CSA

The UK's Child Support Agency is part of the government's Department for Work and Pensions and is responsible for running the child-support system. The role of its officers is to assess family situations and to collect and pay necessary child-support maintenance, ensuring that parents who live apart meet their financial responsibilities to their children.

If a non-resident alleged parent denies his or her parental status, the CSA will discuss the matter with him or her in detail, including the reasons for this denial. They will then inform the person with custody and care of the child (usually, of course, the mother) of the discussions.

In those cases where a child's parentage is disputed prior to a child-maintenance assessment, the CSA can sometimes presume that the alleged non-resident parent is one of the child's parents and assess maintenance on this basis. The circumstances in which this can occur are set out in detail in Section 26(2) of the Child Support Act 1991, summarised as follows:

If the non-resident alleged parent was married to the child's mother at any time between conception and birth of the child, and the child has not since been adopted.

If the non-resident alleged parent is named as the child's father on the birth certificate and the child has not since been adopted.

If a non-resident alleged parent refuses to take a DNA test.

If a non-resident alleged parent has taken a test that shows that there is no reasonable doubt that the alleged non-resident parent is the parent of the child.

If the non-resident alleged parent has adopted the child.

If the court has made an order under Section 30 of the Human Fertilisation and Embryology Act 1990, covering situations where a child is born as a result of certain fertility treatments.

If there has been a declaration of parentage made in a court in England, Wales and Northern Ireland, or a declaration of parentage by a court in Scotland that the non-resident alleged parent is a parent of the child is in force, and the child has not since been adopted.

If the non-resident alleged parent has been found or judged to be the father by a court, in proceedings where parentage was not the central issue.

This presumption of parentage can, of course, be challenged. A non-resident alleged parent can dispute the decision to make a child-maintenance assessment, and he or she can also dispute such an assessment on the grounds of one of the previously listed presumptions. It is up the person challenging the presumption of parentage, not the CSA, to prove that they aren't the actual parent. In all cases, a non-resident alleged parent who disputes parentage must continue to pay child maintenance while tests are being performed. If the dispute is successful, the CSA will pay back any money it has collected on the child or children's behalf.

If the non-resident alleged parent denies his or her parentage and none of the previously listed presumptions apply, the CSA will usually suggest a DNA test. However, both the non-resident alleged parent and the parent with care and custody must agree to take the test and provide samples for testing. The parent with care and custody will also have to give consent for the child or children to be tested.

If care and custody of the child is with a person other than his or her biological parent, and both parents are non-resident, it will be the parents who need to take the test, but it will normally be up to the person with care and custody of the child to give consent for him or her to be tested.

DNA testing might delay the completion of a child-maintenance assessment, but it won't alter the date from which child maintenance is payable. If the non-resident alleged parent is found to be a child's biological parent, he or she will have to pay any child maintenance due, including payments in arrears from the date on which the child-maintenance assessment took effect. He or she will also have to pay the testing fee.

A non-resident alleged parent who refuses to take a DNA test will be presumed to be the parent of the child. The CSA will then work out an appropriate level of child maintenance, which the non-resident alleged parent then has to pay.

In certain cases (for example, with a child conceived via IVF treatment), a DNA test might not be appropriate. In such instances, the CSA might apply to a court, appealing to them to order a determination of parentage. DNA tests that have been commissioned by a non-resident alleged parent without the CSA's involvement might not always be accepted by the agency, who may well order another test to be carried out by an approved laboratory.

For more information, visit the CSA's website at www.csa.gov.uk.
(Source: *www.csa.gov.uk*)

CAFCASS

The Children and Family Court Advisory and Support Service was founded in April 2001 and is a national, non-departmental public body for England and Wales, bringing together services previously provided by the Family Court Welfare Service, the Guardian ad Litem Service and the Children's Division of the Official Solicitor. It is independent of the courts, social services, education and health authorities, as well as of all other similar bodies.

The function of the CAFCASS is to:

- Safeguard and promote the welfare of
  the child;

- Give advice to a court about any application
  made to it in proceedings regarding disputed
  parentage;

- Make provision for children to be
  represented in such proceedings;

- Provide information, advice and support for
  children and their families.

- The court will ask the CAFCASS to help in
  situations where...

  - parents are separating or divorcing and
    have not agreed on arrangements for their
    children;

  - social services have become involved and
    the children might be removed from their
    parents' care for their own safety;

– the children might be adopted.

– Essentially, there are four types of
 CAFCASS officer, as follows:

Children and Family Reporters – These officers
become involved when parents who are
divorcing or separating have been unable to
reach an agreement about arrangements for
their children. Sometimes, agreement can be
reached without having to involve the court any
further, but, if not, the children and family
reporter write a report for the court.

Children's Guardians – These people represent
the interests of the child during cases in which
social services have become involved and in
contested adoptions.

Reporting Officers – These officers ensure that
parents understand what adoption means for
them and their child, and determine whether or
not the parents consent to it.

Guardian ad Litem – These officers are
occasionally appointed by the court in cases

where parents who are divorcing or separating have been unable to reach an agreement concerning their child or children's care and custody, often as a result of there being some particular difficulty in the case. The role of the guardian ad litem is to provide separate representation to argue for the rights and interests of the child.

For more information, visit the CAFCASS website at www.cafcass.gov.uk.
(Source: *www.cafcass.gov.uk*)

8

ACCESS AND GUARDIANSHIP

PARENTAL RESPONSIBILITY

Unlike mothers, fathers don't always have what's known as parental responsibility for their children. Today, more than one in three children in the UK are born outside marriage, and many parents are unclear about who has legal parental responsibility for their child or children.

'Parental responsibility' is the right to make important decisions about a child's life in areas such as medical treatment and schooling. According to the current law, a mother always has parental responsibility for her child, but this responsibility is bestowed on to the father only if he is married to the mother or has acquired legal responsibility for his children.

Unmarried fathers can acquire parental responsibility for their children in three different ways, depending upon when their children were born. If their children were born before December 2003, unmarried fathers can obtain parental responsibility by either:

- marrying the mother of their child, or registering a parental-responsibility agreement at a court or applying to a court.

- If their children were born after December 2003, unmarried fathers can obtain parental responsibility by either:

- registering the child's birth jointly with the mother, or
  - marrying the mother of their child, or
  - registering parental responsibility at a court.

There are also other ways in which a father can obtain parental responsibility, as follows:

- By drawing up with the mother a parental responsibility agreement, which must then be signed by both parents and lodged with a court.

- A court can make a Parental Responsibility Order if the parents cannot agree on the father having parental responsibility.

- It's possible for more than one person to have legally recognised parental responsibility for the same child at the same time. In such cases, the parental responsibility is shared by each such individual, but each can act alone without reference to the other in order to safeguard and protect the child.

CONTACT AND RESIDENCE

These criteria are covered in Section 8 of the Children Act 1989 and are agreed between private individuals. In cases where someone seeks an order under Section 8 in respect of a child who is in the care of a local authority, it will be considered a matter for public law and rules relating to children's guardians will apply.

In private-law cases, however, the child isn't

automatically a party to the proceedings and won't automatically be represented by a guardian. However, a court can request a welfare report under Section 7 of the Children Act, either from a local authority or from a children-and-family reporter from the CAFCASS. The report will usually inform the court of the child's wishes and feelings, but the officer will recommend what they think is in the child's best interests, in the circumstances of the case, rather than just advocate the child's wishes.

In some circumstances, the court might order that the child does become a party to the proceedings. At this point, both a solicitor and a children's guardian (again, an officer of the CAFCASS) are appointed to represent the child in the proceedings. As in public-law proceedings, if the child and guardian fail to agree on what recommendations to make to the court and the child is of sufficient age and understanding, they will be able to instruct a solicitor directly to represent their views, and the guardian will present his or her own views to the court.

- Certain people can make an application for a residence or contact order under Section 8, as of right, including the following:

- The parent or guardian of a child;

- Anyone who holds a residence order in respect of that child;

- A married step-parent of the child in situations where the child lived with the step-parent as a child of the family;

- Anyone with whom the child has lived for at least three years (a period that need not have been continuous but must have been recent);

- Anyone who:
  - has the consent of everyone involved and who holds a residence order already in place, or
  - has the consent of the local authority where the child is in their care, or
  - has the consent of everyone who has parental responsibility for the child.

If an applicant is unable to apply for the order as of right, he or she can make an application to the court for leave to issue the application. In deciding whether to grant such leave, the court will consider, amongst other things:

• The nature of the application;

• The applicant's connection with the child;

• The level of risk there might be of the proposed application disrupting the child's life to such an extent that they should be harmed by it.

In this way, more distant family members, such as grandparents, are able to seek orders in respect of their grandchildren.

THE WELFARE CHECKLIST
THE CHILDREN ACT 1989, SECTION 1

When a court considers any question relating to the upbringing of the child under the Children Act 1989, the court must have regard to the welfare checklist set out in Section 1 of that Act. Among the things that the court must consider are:

- The ascertainable wishes and feelings of the child concerned (considered in light of his or her age and understanding)

- His or her physical, emotional and/or educational needs

- The likely effect on the child of any change in his or her circumstances

- The child's age, sex, background and any characteristics of his or hers that the court considers relevant

- Any harm which the child has suffered or is at risk of suffering

- How capable each of the child's parents is at meeting the child's needs, along with the level of such capability of any other person in relation to whom the court considers the question to be relevant

- The range of powers available to the court under the Children Act 1989 in the proceedings in question

For all proceedings under the Children Act 1989, when a court considers a question of the child's upbringing, the child's welfare is the court's paramount consideration.

RESIDENCE ORDERS
THE CHILDREN ACT 1989, SECTION 8

These orders determine where the child is to live, and with whom. The granting of a Residence Order to someone automatically gives him or her parental responsibility for the child. Parental responsibility obtained by virtue of a Residence Order will continue until the order ceases.

A Residence Order lasts until the child is sixteen years old, unless the circumstances of the case are exceptional and the court has ordered that it should continue for longer. Such an order might also be granted to more than one person, and can be made jointly to an unmarried couple.

Residence Orders prevent anyone from changing the surname of, or from removing from the UK for more than one month, any child who is the subject of the order, without the prior agreement of everyone with parental responsibility or an order of the court.

CONTACT ORDERS
THE CHILDREN ACT 1989, SECTION 8

Contact Orders are those that require the person with whom a child lives to allow that child to visit, stay or have contact with a person named in the order. Such orders continue until the child is sixteen years, after which age the court will make Contact Orders in only exceptional circumstances.

The exact nature of the contact specified in a Contact Order can either be direct (eg face-to-face meetings with a person) or indirect (eg correspondence by letter, video, exchange of Christmas cards, etc).

Some Contact Orders are very specific in their references to times, dates and arrangements for contact, while others are more open, leaving the detail of arrangements to be made by the parties involved.

Parents seeking contact with their children can obtain Contact Orders, but they can also be drawn up to stipulate levels of contact between siblings or between the child and wider family members.

Sometimes a Contact Order will give directions that the contact is to be supervised by a third person. An order might also only be actionable for a specific period or contain provisions that operate for only a specific period.

Contact Orders are, of course, orders issued by a court, and so failure to comply with their terms might be viewed as a contempt of that court and might incur serious consequences.

For more information, visit www.direct.gov.uk/audiences/parents. (Source: *www.cafcass.gov.uk* and *www.direct.gov.uk*)

9

AN EXCELLENT
SOURCE OF SUPPORT

1974 saw the establishment of the UK-based
charity Families Need Fathers (not to be
confused with Fathers 4 Justice). Its name,
however, can be misleading and sound biased
towards men, but the members of the FNF
firmly believe that the best parents are both
parents. Over the past thirty years, the charity
has been providing information and support to
parents of both sexes, whether married or not,
and is chiefly concerned with the problems of
maintaining a child's relationship with both
parents and, indeed, all relations during a
family break-up. Over the past few years, the
volunteers at the FNF have dealt with an
increasing number of disputes concerning

access for grandparents and have seen a rapidly increasing surge in female membership.

Jim Parton, a spokesman for the FNF, defines the charity's goal as that of achieving a 'presumption of equality' after a relationship between two parents has deteriorated, meaning that, instead of the current situation, where custody and parental responsibility is generally awarded automatically to the mother, both parents should be given equal rights under the law. The charity's objectives – which are set out fully on their website at www.fnf.org.uk – are as follows:

• To provide relief for parents and their children and other close family members suffering from the consequences of divorce or separation by providing advice, assistance and other support and, in so doing, helping parents stay in touch with their children after divorce or separation;

• To further the emotional development of children whose parents have divorced or separated by encouraging shared-parenting arrangements, which enable such children to have continuing and meaningful relationships with both their parents.

- To conduct study and research into: the problems of children who are deprived of the presence of a parent in their families, and the problems involved in establishing good relations between parents living apart from their children, and to publish the useful results of all such study and research in order to encourage appropriate changes in professional and public opinion;

- To relieve parents in poverty by helping to obtain and promote the provision of free legal advice, assistance and other free legal services that such persons cannot access because of their lack of financial means.

In terms of DNA-based paternity testing, the members of the FNF believe that developing technology will result in the testing procedures becoming increasingly straightforward. They feel most strongly that accredited laboratories, rather than individuals, are the appropriate forums through which reliability should be obtained.

However, as Jim Parton, their spokesman, warns, 'The truth can be your enemy.' The volunteers at the FNF have encountered many situations in which assumed fathers have received negative results to DNA tests, and in these cases it's rare – regardless of how much time has been invested – for a court to award a Parental Responsibility Order in favour of the person whose paternity has been excluded. In such cases, the child might lose access to both a loving parent and, in all probability, an entire support network of extended family. The FNF believes that such cases need to be reviewed, and that access should be granted to a man who has acted as a father during a child's lifetime.

Meanwhile, the FNF can advise individuals on how to obtain a DNA sample and where to go for the testing, but ultimately the decision is a matter for each individual concerned.

In cases where parents have been married and a 'presumption of paternity' is held, the FNF believes that there should be a statutory presumption of paternity that holds true unless the father disputes paternity. As their members see it:

'Marriages do not break up in one day. Often there are difficulties over years with spouses reconciling after trial separations. During some of these periods, our experience is that it is not uncommon for wives to have relationships with other men. If the original couple reconcile, the husband might bring up another man's child for many years, possibly unbeknownst to him and his wife.

'In cases where an attempted reconciliation fails, it is not uncommon, in our experience, for a wife to provoke or attempt to hurt her husband with taunts that he is not the father of their children. He needs to know the truth.

'We do not believe, however, that there should be a presumption if the husband requires proof.'

The FNF is staffed with volunteers experienced in all areas of family law who can help putative parents in their quest for truth, but, as a charitable organisation, it relies heavily on donations gathered through membership. Its website, at www.fnf.org.uk, provides details of how they can help people wanting to determine the truth about their relationships with their family members. With the majority of its volunteers having had first-hand experience of situations concerning disputed paternity, the charity's emotional support is both priceless and freely given.

10

GENERAL INFORMATION

GROWTH OF THE MARKET, STATISTICS AND A LIST OF RECOMMENDED DNA-TESTING COMPANIES.

According to the office of National Statistics:

'There has been a decline in the proportion of families headed by a married or cohabiting couple and a corresponding increase in the proportion headed by a lone parent. In 2002, 73 per cent of families in Great Britain consisted of a married or cohabiting couple and their dependent children. This is a proportion that has declined steadily since 1971, when 92 per cent of families were of this type.

'The large growth in the proportion of lone-parent families (from 8 per cent of families in 1971 to over a quarter of families [27 per cent] in 2002) has mainly been among families headed by a lone mother. Lone-father families have accounted for 1 to 3 per cent of families since 1971, whereas the percentage of lone-mother families has risen from 7 per cent in 1971 to 24 per cent in 2002.

The percentage of families headed by mothers who have never married (ie single) has risen from 1 per cent in 1971 to 12 per cent in 2002. The percentage of families headed by mothers who were previously married and are now divorced, widowed or separated has risen from 6 per cent t0 12 per cent during the same period.'

With this being the case, with many more children being born into unstable relationships, the question of paternity is today more important than ever. That aside, paternity of children is confirmed six times out of seven.

The growing paternity-testing industry in the UK is said to be worth ten times its value of a decade ago, now estimated as being around £60 million per annum, a figure that is predicted to rise to over £100 million per annum by 2008. Resources inside the industry estimate that there are some 30,000 tests taking place each year, and the number is rapidly increasing.

In 2002, I initiated some market research to find out why the number of people seeking paternity tests is on the increase. Here's an extract of my findings:

Key demographic factors that will affect the increased demand in testing:

• Births outside marriage grew by 11.25 per cent in the decade between 1991 and 2001. This is important evidence to support the growth figures for the paternity-testing market.

• There is evidence to suggest that attitudes among younger unmarried fathers to paternity tests are radically different to those of the older generation and married men. A survey reported in the *Scottish Daily Record* that 25 per cent of men under the age of twenty-one said that they would demand a paternity test if a woman claimed they had got her pregnant.

- In 2001, there were 238,886 births outside marriage and around 24,000 births by women younger than eighteen. The problem of paternity among the young is evidently a large and growing issue. The birth rates outside marriage, meanwhile, continue to decline.

- In 2001 there were 143,818 divorces granted in England and Wales, compared to 141,135 in 2000 – an increase of 1.9 per cent, the first rise in the number of divorces per year since 1996.

- 70 per cent of all divorces in 2001 were between couples where the marriage had been the first for both parties, compared with 80 per cent in 1982, making for more second families and more complex family structures.

- In 2001, a total of 146,914 children under sixteen years old were in families where the parents divorced, and just under a quarter of these children were aged under five. This points to substantial growth in the number of children who might grow up with paternity issues.

- In the early 1970s, less than one in twelve of all families with dependent children was a lone-mother family. By 2000, this proportion had almost trebled to just under one in four. More single mothers by definition means more absent fathers.

- Single lone-mother families – where the lone mothers have never married – continued to grow between 1996 and 2000 until they made up one in nine of all families with dependent children. Dependent children in single lone-mother families form one in eleven of all dependent children.

- Single lone mothers form two in every five lone parents, while over one-third of all dependent children in lone-parent families live in single lone-mother families.

- Over one-third of all lone mothers reported that they had lived in an informal union that had ended in ways other than marriage. The corresponding proportion for married women with dependent children is only one in ten.

The following points all serve to indicate a move to shorter-lived, more informal, temporary relationships between child-bearing couples, all of which criteria are likely to boost the frequency with which paternity issues are raised.

With cohabitation among non-married people aged sixteen to fifty-nine running at an average of 24 per cent across the UK, when broken down regionally, this figure presents some interesting regional variations, as demonstrated in the chart below. For instance in the East Midlands, co-habitation is 9 per cent more popular than in the Northwest.

Percentage of non-married people aged 16-59 cohabiting

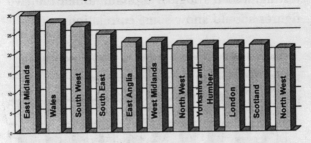

'Within Marriage/Outside Marriage'

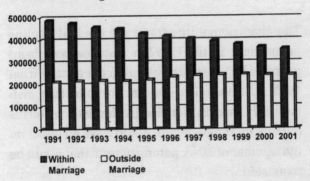

■ Within   □ Outside
Marriage   Marriage

The graph above illustrates the declining rate of
children born within marriage in comparison
with those born outside marriage. The number
of children born outside marriage, for instance,
has increased by 11.25 per cent in the past
decade. This rise is concurrent with the
increases seen in the demand for paternity

testing, and it's logical to assume that the two figures should show some correlation.

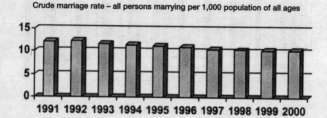

Crude marriage rate – all persons marrying per 1,000 population of all ages

The chart above illustrates the gradual decline in marriage in the UK. For instance, marriages dropped by 2.8 per thousand in the decade from 1991 to 2000. The decline in marriage will naturally feed into a rise in the level of births outside marriage, which could have an effect on the number of DNA paternity tests that might be requested.

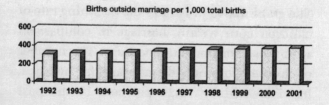

Births outside marriage per 1,000 total births

As the chart above illustrates, the number of births outside marriage increased from 30 per cent to 40 per cent of total births during the 1990s, largely due to fewer couples getting married and more couples having children outside or before marriage.

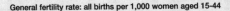

General fertility rate: all births per 1,000 women aged 15-44

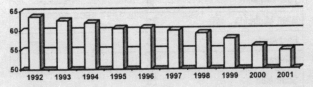

The above chart, however, illustrates that there is a continuing decline in the overall level of births – indeed, a 0.9 per cent decrease in general fertility rate. Overall, the number of couples having children is declining, which means that the absolute size of the market for baby- and child-focused products and services will level out and decline in the medium to long term.

COHABITATION AMONGST NON-MARRIED PEOPLE AGED SIXTEEN TO FIFTY-NINE, 1998–2001

LOCATION IN UK	PERCENTAGE
Northeast	22
Northwest	21
Yorkshire and the Humber	22
East Midlands	30
West Midlands	23
East Anglia	23
London	22
Southeast	25
Southwest	27
England	24
Wales	28
Scotland	22

(Source: *General Household Survey*, Office for National Statistics; *Continuous Household Survey*, Northern Ireland Statistics and Research Agency)

REAL-LIFE EXAMPLES AND RESULTS OF DNA TESTING

Test Type: **Paternity**
Test Result: **Negative**

CASE STUDY: JACK BAILEY, LONDON

Jack is your average bachelor; he enjoys going out with his friends and living the single life. On one particular night two years ago, he met up with Jenny, a girl who also enjoyed her freedom and time with friends. Jenny became a part of Jack's extended social circle and eventually got together with Max. Jenny was pretty, funny and outgoing, but her time with Jack was never a real relationship; in fact, it was practically a one-night stand.

One night, when Jack was out with his

friends, he heard a rumour that Jenny was pregnant and that she was telling everyone that he was the father. Jack refused to believe the gossip because he knew that she had been in a long-term relationship for seven years, even when they were together. He was convinced that their one night together could never have produced a child. Jenny, however, insisted that the child was his and that she hadn't been intimate with her long-term partner at the time of conception.

Despite her allegations, Jack didn't contact her, and for nine months he had only vague ideas of what had become of her and the baby, except what went around among their friends. During that time, Jenny had started dating someone else. The new boyfriend, having been convinced by Jenny's claims that the child was Jack's, began to harass Jack. He told Jack that he wasn't doing anything for the child and that he should be providing financial assistance.

Still convinced that he wasn't the father, Jack received a call from Jenny, who told him that she had ordered a DNA test on his behalf and wanted him to take it and acknowledge that the child was his. Jack, knowing that he had nothing to hide, agreed and took the test. He had been made out to be the bad guy by his friends, his family and even those he didn't know, especially because, since Jenny's pregnancy and the baby's birth, he'd had nothing to do with her. Now, he was ready to prove that the baby wasn't his and be done with the matter forever.

When the report was delivered, Jack and Jenny sat down at Jenny's parents' house and opened the envelope. Jenny's parents hovered in the next room, and Jack felt very uncomfortable. It occurred to him that perhaps Jenny had waited so long to have the test done because she had been hoping that he would finally cave in under all the pressure and admit that the child was his. She had been convincing herself that the baby was his this whole time because she wanted it to be true so badly. Now the moment of truth had arrived. There was nothing more to do but open the letter and see the result.

As the letter was read, Jack and Jenny had very different reactions – Jenny's one of disbelief and tears and Jack's one of pure elation. Jenny sat in complete silence. She had no response. Jack, on the other hand, was 'really happy'. The test had, of course, come back negative. Jack tried to comfort Jenny, but knew there wasn't much he could – or, indeed, wanted – to do. Jenny's child had nothing to do with him now.

And what will he do now? 'I have no reason to be ashamed,' he affirms. 'These things happen. I was confident the whole time the baby wasn't mine, but she disagreed. I feel sorry for her, but I don't want anything more to do with her. I'm going to keep loving my life as I'm loving it.'

Test Type: **Paternity**
Test Result: **Positive**

CASE STUDY: CARLY, BRISTOL

Carly and her partner were looking forward to starting a new life together after he had just come out of a long relationship that had produced a young son, John. John looked just like his father and they were close in every way.

About a year later, John's mother came around, six months' pregnant, and declared that the baby to be was the daughter of Carly's partner. Prepared to take full responsibility, Carly and her partner were ready to aid John and his sister-to-be in every way possible.

Six years passed and Rebecca, the children's mother, found herself in a difficult situation, having been kicked out of her house and declared unfit to raise the children until she could establish a stable environment for them. John was happy to move in with Carly and his father, as their relationship had grown very close. However, although Carly and her partner had always tried to treat Chrissie as their daughter, she was very resistant to the change and insisted on staying with her mother. Carly and her partner decided to fight for full custody of John, but Rebecca wouldn't let him go unless it could be proved that both John and Chrissie were, in fact, his children.

In July 2005, Carly and her partner filed for custody and John was able to live with them for the time being. Rebecca had previously got a court order against the children's father, making it quite difficult for the case to progress. To put these issues to rest, Carly and her partner decided to have a DNA test done. They were convinced that the test for Chrissie would come back negative because they had no bond with her to speak of, the mother often discouraged the familial connection and they looked entirely different from one other.

To their surprise and disappointment, the results came back positive: Chrissie's DNA profile bore a marked resemblance to her father's. 'It was absolutely devastating,' Carly recalls. 'We were completely taken by surprise!' The couple obviously believed in the validity of the results but felt very disconcerted with the new reality that faced them.

Carly and her partner attempted to accept the unexpected news and carry on like a family, but their relationship consequently suffered as they were now faced with another child to deal with and extended the family that Carly desired to have with her partner. There were new levels of stress and resentment to deal with, too, and important family matters were put on hold. They were also plagued with added worries of more children and broken relationships.

Had the results been different, they believe that they could have cut their ties with Chrissie and focused entirely on John, who had during this time become somewhat lost in the confusion and shuffling of court papers and tests. His relationship with his own mother and sister suffered likewise, as there was no real bond between them. At times, the awkwardness in the connection between mother and son and between father and daughter became unbearable, resulting in strong feelings of anger, hatred and resentment on all sides. 'Instead of feeling like she was a daughter, Chrissie felt more like a child that was coming over for afternoon tea,' Carly explains. 'I couldn't help but think, "If it weren't for you, we would be moving in a different direction and things would be different." I know it's not her fault, and I don't mean to do it; those are just the feelings I sometimes have.'

Carly admits further that if, in the end, Chrissie didn't want to be a part of their family any longer, there would be more relief than sadness because it would mean that she and her partner could finally move forward with their lives instead of putting up barriers, controlling unnecessary contention and limiting interaction with certain family members.

'At the end of the day, it's nobody's fault,' says Carly of their situation, although she believes that mothers and fathers do have a responsibility to know who the rightful parents are of their children. 'It's not fair to the children,' she explains, 'as it puts them in difficult situations.' Overall, Carly is happy that she and her partner had the DNA test done, for it eliminated all doubts for them, but it does stand as a warning that not all results are those that are expected.

In spite of all the hard feelings that have resulted from this test, there has been some amazing good to come of it as well. Carly reports that John is doing extremely well by being in a family that truly loves and encourages him to be his best. 'People that have known him through this all say that he has never looked better. He has more colour in his face and seems like such a happier child.' Carly and her partner have taken great care to make him a top priority and to provide an environment in which he can continue to grow up healthily and happily. As for Chrissie, Carly wishes her the same success, and yet, without that special bond, who knows if it will ever be accomplished through them?

Test Type: **Twin-Zygosity Test**
Test Result: **Positive (Identical Twins)**

CASE STUDY: JEMMA, TUNBRIDGE WELLS

No one knows better than Jemma Barrow the difficulties involved in raising twins. Everything takes twice as long and requires double the energy. Nevertheless, her two boys were a blessing in her life and she wouldn't trade all the runny noses or late nights for anything in the world; she loved her babies and felt so lucky to be their mother.

Going out with the twins was an adventure in itself. People who Jemma didn't even know would come up to greet her and share their condolences or excitement with her. A common question she was asked was 'Are they identical or fraternal twins?' In actuality, Jemma couldn't answer, as they'd never thought about it much themselves. The boys looked very alike, but she'd heard that it didn't necessarily mean that they were identical. She found that she wanted to give people a correct answer instead of just guessing. The strangers, meanwhile, felt it quite appropriate to tell her exactly why they thought they were identical or not.

As the years passed, one of her sons started to have something of an identity crisis. He was only five years old at the time but was growing tired of being a twin. He yearned for his individuality and possibly felt that it couldn't be achieved completely as a twin. Jemma and her husband sympathised with their son but didn't know what to tell him. They hoped that, through taking a DNA test, they might be able to settle the ongoing struggle their son was having and finally be able to put the matter to rest.

Jemma and her husband waited for the test results to come through, hoping that they would show that the twins were non-identical, in which case they would be able to tell their sons that, even though they looked so much alike, they would eventually become more distinct from each other in their behaviour and even start to look different. If it turned out that they were identical, however, Jemma and her husband hoped that the fact that they shared so many wonderful qualities between them would help create a stronger bond and, at least, allow them finally to know the truth.

When the test results were delivered, they indicated that the twins were identical. 'We were hoping for non-identical, but it's nice to know,' said Jemma, admitting further that the test was really just for peace of mind and that, when it came down to it, the results really didn't matter. 'We just wanted to know what to tell people when they asked. Now we can say assuredly that they are identical.' She and her husband hope that their son will be able to deal with the news and that no other crises will follow, as she has enough to worry about with twins as it is.

Test Type: **Paternity for Immigration Purposes**
Test Result: **Negative**

CASE STUDY: ADE, LONDON

Ade Ayemeni was very surprised when his girlfriend told him that she was pregnant for the second time, after the couple had had one child together, so he asked her to confirm that she was pregnant and that she was certain about the due date. He'd been away for a while, so the dates puzzled him.

At the time, Ade and his girlfriend were living separately in Nigeria and he continued to worry about the paternity of the unborn twins throughout the pregnancy. He was still at college and felt the responsibilities of bringing up three children bearing down on him. He wasn't financially stable, and he and his girlfriend hadn't discussed settling down together and having more children.

Ade's doubts over the unborn children's paternity continued to nag at him, and when the twin girls were born he felt no surer that he was the father than he had nine months earlier. Nevertheless, his girlfriend continued to tell him that the twins were his.

Ade decided that the only explanation for the twins' conception was that his girlfriend had been unfaithful, and for the first eighteen months of the twins' lives he stayed away while he grappled with raising the money to support two more children who he believed weren't his. During this time, he talked to friends and family members who asked if his girlfriend had been seeing someone else. His girlfriend continued to protest that Ade was the father, and eventually he was convinced that this was indeed the case.

From that moment on, Ade brought up the three girls with all the love that he could give, funding their education and making sure they wanted for nothing. Then he left Nigeria to seek his fortune in the UK, intending to bring over his three girls when he was settled. While he was there, however, he met and fell in love with a British lady. They were married and had three boys.

Eleven years after the wedding, Ade submitted visa papers that would enable him to bring over his girls from Nigeria, and he was advised by his solicitor to organise a DNA test for the girls so that he could prove they were his.

The results of the DNA test turned Ade's world upside down, and he was devastated to learn that he wasn't the father of the twins after all. He loved all of his three girls equally but felt deeply betrayed by their mother, who had been lying to him for sixteen years.

Very soon after this, the three girls arrived in the UK and settled into life with their father. Ade explained the situation to all of his children: that he loved the twins and that, to him, they were his daughters and he loved them no less now than he had before.

Ade lived with this news for three years until he decided to travel back to Nigeria to confront the girl's mother and find out the truth about why she had lied to him. The visit, however, was very unfulfilling; when they met, she continued to deny that she'd been unfaithful, even though he presented her with irrefutable DNA evidence.

Meanwhile, Ade is committed to his three girls and is continuing to give them financial support for their education in the UK. He hopes that the twins will always love him as their dad. He is a proud father and wants only the best for all of his six children.

Test Type: **Ethnicity**
Test Results: **87 per cent Northern European, 8 per cent Native American and 5 per cent sub-Saharan descent**

CASE STUDY: SAMANTHA, LONDON

Every person has a unique story, but not every person's starts out like Samantha's. 'I was a black-market baby,' she confesses. While this term conjures up scenes of children being sold in marketplaces and so on, 'black market' isn't exactly what she means.

Seventy-two-year-old Samantha was born into an interesting family. Her mother was an abusive alcoholic and her father, although wonderful, cowered under his wife's hostility and aggressiveness. During one incident in Samantha's seventeenth year, while her mother was in yet another of her alcoholic rages, in between belittling and swearing at her two daughters she confessed that she hated them and that they had been adopted. Samantha was shocked and traumatised by her mother's admission but in fact remembered very little of that part of her life. 'I just blocked in out and it's all very fuzzy in my head,' she confesses.

From that point onwards, Samantha put a lot of effort into finding out who she was. Her mother finally told one of her daughters that she was actually the daughter of her sister, but she would never tell Samantha who that sister was. The only thing that Samantha knew was that her parents had split for a short while and then got back together at about the same time that Samantha had been conceived and born. Her mother refused to say more and her father said that he couldn't discuss the matter because he'd 'made a promise'.

For years, Samantha lived with the agony of not knowing the true details of her parentage.

She hoped that on their deathbeds her parents might tell her the whole story and give her some type of clue, but this was also to no avail. Her father had been sick for some time, but, when he got worse and Samantha prepared to visit him, her mother said that he was actually looking and feeling better and not to worry about making the long trip to see him. Only days later, Samantha called to see how her father was and was told by her mother that he'd died on the night Samantha had planned to visit. Samantha never even got to say goodbye. She was very angry that her mother denied her ever knowing the information she so desperately wanted, but she crossed her fingers and hoped that, perhaps in some act of mercy, she might one day tell her who she really was.

No such luck. Years after her father's death, her mother fell ill and was near death herself. Even in her extremely weakened state, however, she refused to tell Samantha who her parents were. Then she died, leaving Samantha to uncover her true identity alone.

After the death of her mother, Samantha located her birth certificate, which helped her to discover new information. Further enquiries led her to the hospital in which she'd been born, and she also contacted the doctor who'd delivered her, who informed her that her real mother had entered the hospital under the guise of her adoptive mother. The doctor was very brief and, according to Samantha, 'quite nasty' about the whole ordeal; he, too, had sworn that he would never tell what had really happened and ended their discussion by saying, 'The case is closed!'

And so, with all other avenues closed, Samantha turned to DNA testing, hoping that an ethnicity test would prove her relationship to her father. She had meanwhile come up with her own explanation to who she was, figuring that, during the time when her parents had separated, her father had had an affair with another woman, who, instead of keeping the baby for herself, gave it to Samantha's father and wife to raise. Samantha had asked her father if he was her real father, but he'd never given her a straight answer. She hoped that taking a DNA test would open up new lines of enquiry.

Her test results confirmed that Samantha was of 87 per cent Northern European, 8 per cent

Native American and 5 per cent sub-Saharan descent Now she is faced with a lot of genealogy work and the task of seeking out possible relatives. Regardless of dead end after dead end, however, Samantha remains extremely hopeful and positive. 'I am the start of a new generation,' she exclaims. 'These results have given me a renewed hope that I will someday find out who I am.'

Test Type: **Paternity**
Test Result: **Negative**

CASE STUDY: KIRSTEN, BRAINTREE

After a seven-year relationship, Theo and Laura, once a happy couple, found reason to file for divorce. They had three small children who Theo loved and cared for very much, and he was still fully committed to the needs of each one of them, despite the split.

After the divorce, Theo met a wonderful woman named Kirsten with whom he started a new life and a family with the birth of their little girl, Sarah. By now, Theo's own children had grown up and his oldest, Shayne, had reached fourteen years old. Around this time,

Theo started hearing rumours that his ex-wife had had numerous lovers and affairs and subsequently borne children that were not Theo's. For six months, these rumours circulated in Theo's head, and his constant dwelling on them was starting to affect him in many ways. His work and relationship with Kirsten suffered from his stress and obsession over what he'd heard, and it was as if he was unable to move forward with his life. Distressed but determined to prove that the rumours were groundless, Theo decided that a DNA test would ease his mind, quieten all that talk and prove that he really was the father of his children.

Laura initially refused to take the test, claiming that it was unnecessary and ridiculous, but finally allowed for at least Shayne to be tested. They sent Shayne's swabs to the lab and anxiously awaited the results.

When they were delivered a week later, the results were heartbreaking and changed Theo and Laura's lives entirely. Theo, it turned out, wasn't Shayne's father. He was unconvinced, however, and asked Laura to let him do a retest, but she wouldn't hear of it. To test the validity of the results, Theo and Kirsten did a paternity test on their daughter, Sarah, which came back positive. No matter how he begged her, however, Laura wouldn't let Theo do any further testing and has even denied Theo contact with the children. He also pleaded with her not to tell the children, but Laura went ahead and told them without any explanation.

'It's been crazy,' says Kirsten of the whole ordeal. 'Theo is heartbroken and feels so betrayed. He loves his children, and it's maddening because he hasn't done anything wrong yet he's receiving the brunt of it all.' Not just Theo; according to Kirsten, the whole affair has been rather hard on the children. Their mother explained hardly anything, and they're starting to experience feelings of abandonment. 'It's been very traumatising for them,' says Kirsten, 'and their mother won't even let Theo meet with them to explain. Theo did the test just to appease others, but it didn't come out the way he thought it would.'

Theo and Kirsten are praying that they'll eventually get to see the kids again and that hopefully they can become part of their father's life once more. 'He is too emotionally tied to those kids. It's not like you can turn off fourteen years like it's a switch or something,' say Kirsten. However, it appears that that is exactly what Laura wishes Theo would do and hasn't relented in her unwillingness to co-operate.

The one good thing to have come of the experience is that Theo and Kirsten have grown considerably closer. 'Through support and understanding, our relationship has grown stronger... but given the chance again, I don't think he would have done it,' says Kirsten. 'Once you reach that point, there's no turning back. You think you're getting one thing but you might be getting another. All he wanted to do was prove a point and it didn't prove anything.'

Test Type: **Paternity**
Test Result: **Negative**

CASE STUDY: GARY, LONDON

In September 2000, Gary met a wonderful woman named Abigail. Despite the fact that they lived 750 miles apart, they formed a fast friendship that eventually became a romantic relationship. For nine months they travelled every weekend to be together and truly enjoyed each other's company. However, as Gary's business began to grow, heavy work responsibilities and increased travelling began to take their toll on the relationship and problems started to arise. To ease the stress, Gary and Abigail decided to end things, but they still remained close friends.

After about three months had passed, Gary found that he was really missing Abigail and called to see how she was. He was shocked when she told him that she was pregnant and couldn't figure out what she was telling him exactly. Could the baby be his? Abigail was admittedly unsure because, two weeks after the break-up, she'd attended a party where, feeling sad and depressed, she'd got extremely drunk and ended up sleeping with a man. There was added confusion when the doctor confirmed that she was ten weeks pregnant while she and Gary had been apart for twelve.

As matters unfolded, Gary and Abigail began to communicate more frequently. Gary felt that this contact was important as it could mean more for the future if the baby were indeed his. He and Abigail spent a lot of time on the phone together throughout her pregnancy, and about a month after the baby was born Gary found himself attending business meetings in her area.

Gary was happy to be near Abigail again and called her to arrange a time to meet. She gladly accepted the visit and they found time to get together. As Gary looked at his beautiful daughter, thoughts came back to him of what he had felt throughout Abigail's pregnancy: 'This child could be mine.' They took this time together to talk about the possibility of getting a DNA test done, just to be sure.

Between the time of talking about the test and actually having it done, Gary and Abigail got back together. No matter what the results, Gary decided, his love for Abigail and the baby were unconditional. The results wouldn't determine his devotion to them.

When the results were delivered, Gary and Abigail sat down together, not wanting to open their envelopes. They'd done this all together and hoped that perhaps the information contained within the envelope would allow them to share something more – something even bigger – together. They had long anticipated the news and felt great fear and anxiety, as it was possible that the test would determine the direction in which their relationship would go. Abigail desperately wanted Gary to be the father of her daughter. She truly loved him and couldn't bear to think that the baby might actually have been the result of a one-night stand, a moment of weakness. Gary, likewise, had become quite attached to the girl and felt for once in his life that he might just be willing to give up his life of bachelorhood and be a father.

Unfortunately, these hopes were not realised for the test results were negative. Nevertheless, Gary was true to his word and stayed with Abigail for another two years. Once again, however, the long distance between them made their relationship very difficult and imposed a great deal of stress. Their lives began to go in different directions and it became clear that the relationship couldn't last much longer.

Things eventually ended between Gary and Abigail, but certainly not on a sour note. 'We have remained the best of friends,' says Gary, who looks back on his relationship with Abigail with much fondness and love. He states further that everything he went through with Abigail and the baby really acted as a catalyst for what he wants for his future: 'I never wanted to be married or a father, but everything that happened with Abigail and spending time with the baby hit me like a ton of bricks. I now feel open to different things. I see people who are married and having children and how fast their lives are moving along and I think that this might just be something that I want too.'

Test Type: **Paternity**
Test Result: **Positive**

CASE STUDY: DAVID, DARTFORD

David was extremely surprised when he heard that a woman he'd known twenty-eight years ago was making strenuous efforts to get in contact with him once again. It had to be important, so he contacted her. What she had to tell him turned out to be more than he'd bargained for.

Elizabeth and David had had a brief relationship many years earlier that wasn't particularly fantastic for either of them. Deciding that it would be better to end things, she'd headed back to Ireland while he'd stayed in England. However, during their short time together, Elizabeth had also been spending time with another man, as David found out eighteen months after they'd split up when he discovered that Elizabeth had written to a friend and told her that she'd conceived a child during her time with both men and had borne a daughter. The letter neither indicated who the father was nor asked that someone take up child-support responsibilities; it was merely a letter of information, and David didn't know quite what to make of it.

Feeling somewhat threatened at the thought of having a daughter, David felt that, since Elizabeth was unable to confirm the details of her pregnancy and made no attempt to resolve the problem, he would do likewise. He left matters as they stood.

Five years later, Elizabeth found herself nearby and contacted David again. He agreed to meet her and her daughter, Carrie-Ann, who he was surprised to find had grown up into a charming young girl. The meeting was pleasant and Elizabeth told David that she was actually with another man at the time.

Once again, nothing was said about the possibility that David could be the father. In truth, he didn't feel emotionally mature enough to take on the responsibilities of fatherhood, even though it was still unclear what Elizabeth expected of him. He didn't know what to do, and so, as before, he did nothing and lost contact with her until she phoned over twenty years later.

This time, Elizabeth claimed that she was doing family research and that she felt it would be good for Carrie-Ann to know who her father was. David had been wondering similarly and had begun to feel somewhat distressed over the previous few years at the thought that he'd never done anything to know if Carrie-Ann was actually his daughter. As it turned out, the other man with whom Elizabeth had been involved nearly thirty years ago had remained a family friend and later married someone else. During their time together it had become apparent that he was unable to produce children, so David was the prime candidate for being Carrie-Ann's father.

Happy to have regained this contact, David was anticipating meeting with Carrie-Ann and Elizabeth and felt a desire to explain why it was that he hadn't done anything in all these years.

Through several meetings with Carrie-Ann, David found her to be a very lovely young woman. They got along beautifully and formed a great friendship. He felt relieved that there had been no animosity or hard feelings at his lack of action on finding out whether or not he was the father. They decided together that they should take a paternity test sooner rather than later in order to avoid greater disappointment further down the line after having invested so much time and care into one another.

David felt very attached to Carrie-Ann, yet he feared the results that a DNA test might provide. If the test came back negative, it would have served only to confirm that he'd done the right thing by not pressing his involvement from the beginning, but by the same token a positive result would highlight his stupidity in having missed the opportunity to spend time with a wonderful daughter. There was therefore a small part of him that hoped for a negative result, and yet he'd come to admire Carrie-Ann and knew that there were so many things he wanted to do and share with his supposed daughter but felt hesitant to do so before knowing the truth. He wanted and needed that confirmation that they truly were father and daughter.

Despite all the doubts and hesitation David felt about the results, he was elated when the test results came confirming that Carrie-Ann was indeed his daughter. Although nothing had ever been conditional on the test results, David felt that they stood as a genuine and symbolic chance to connect with a daughter he hadn't known for her entire life. They gave him an absolute and clear-cut answer. 'It was a very nice and emotional reality,' he noted.

Test Type: **Paternity**
Test Result: **Positive**

CASE STUDY: GLENN, UK

Glenn had a passion for scuba diving and often holidayed in Jamaica for that reason, and also because he had a friend who had frequent business on the island. On one visit, he met Jenni, a girl who worked for the dive company. They became good friends and eventually started a relationship together. However, the long distance made it difficult for them to have a normal relationship, so they had to content themselves with seeing each other only a few times a year.

In March, only a few months after Glenn had completed a spring dive trip and spent some time with Jenni, she called him up and told him the news: she was pregnant. Considering how little time they'd spent together, Glenn was shocked and suggested that they take a DNA test to confirm the child's parentage. Jenni's reaction of mistrust encouraged Glenn to drop the idea for the moment and just support her as she dealt with the pregnancy.

Many months later, Glenn went to Jamaica to be with Jenni and witness the birth of the baby. Without telling Jenni he was going to, he brought a DNA-testing kit with him, just so he could get some peace of mind. As he and Jenni waited for the birth, Glenn admits that it was a very emotional time for him. Being there with Jenni and knowing that she was possibly carrying his child made him hope for a positive result.

When the baby was born, Glenn took one look at her and thought, 'She's definitely mine.' He was almost convinced, except for the small doubt that had been niggling at him since Jenni had first told him she was expecting and which left him wanting to know the child's paternity for sure. So, just three days after the birth, Glenn took DNA samples from the baby, although his guilt and fear that the baby might not be his caused him to wait nearly three months before mailing them in. He couldn't bear to imagine the look on Jenni's face when he told her what he'd done or to discover that the baby wasn't his. He had likewise grown attached to the thought of having a child and knew that a negative result would be devastating. Despite those potential consequences, however, Glenn knew that having the test done was the only way he'd find that peace of mind he craved.

As Glenn waited for the results, his friends and family were very supportive and rallied around him, giving him the confidence to accept the results, whatever they might be. When they came back positive, he was overjoyed and relieved, although he felt guilty for having done the test without Jenni's consent. He phoned her up to tell her what he had done and was relieved to discover that she wasn't angry and could understand his suspicions.

Jenni still lives in Jamaica but Glenn is in the process of bringing her and their daughter over to the UK.

Test Type: **Paternity**
Test Result: **Positive**

CASE STUDY: GLORIA, LONDON

Gloria and her husband had raised their grandson, Brandon, from childhood. When he was a young man and preparing to enter the army, Brandon received unsettling news from his girlfriend, Julia: she was pregnant. Brandon was quite angry and explained that there was no way that he was ready to be a father yet. Subsequently, Julia had an abortion and the couple continued to go out.

Three months after the abortion, and unbeknownst to Brandon, Julia stopped taking her birth-control pills. By now, Brandon had enrolled in the army and was at home only at weekends, and even these visits weren't regular. During their time away, Julia began to notice strange changes in her body, and after several months it was obvious to everyone that she was once again pregnant.

The news that his girlfriend was pregnant again enraged Brandon and they broke up just before the army sent him to Poland. They had contact while he was away, as Brandon had completely removed himself from the relationship. While he was there, however, he received a call from Julia, who told him that she'd just delivered a baby boy.

A few months later, on his return home, Brandon visited Julia to see her and the baby. But was the baby his? It was true that the baby had Brandon's striking blue eyes, but Gloria told him that he should make sure because being a father was a big responsibility. It was also true that Julia and Brandon had met under less than ideal circumstances, when she'd been living with another man and was pregnant with someone else's baby. These facts alone generated large amounts of mistrust in Brandon's relatives, who urged him to find out for sure.

Julia, however, did not feel that a test was necessary and actually refused to consent to have it taken. Why should they need a test, she argued, if she already knew that Brandon was the father? Brandon's family took this as a sign that Julia was experiencing doubts, believing

131

that she was worried she would lose Brandon if it turned out that the baby – the only thing keeping them together – wasn't his. Julia's reluctance to consent to the test encouraged thoughts that perhaps she'd been lying about Brandon's paternity this whole time.

Eventually, however, Julia consented to the test. 'It could be that Julia was 95 per cent sure that the baby was Brandon's,' Gloria observes, 'but you can't afford to have an element of doubt where a child is concerned.' Nevertheless, Julia's inclinations turned out to be right and the test results came back positive.

Gloria has doubted Julia's intentions from the beginning and notes how sad it is that her grandson has become the object of someone else's manipulation, a means to an end. 'Brandon is young and impressionable and he wants a family life,' she notes. 'Whether that will end up being with Julia, I don't know. He is very affectionate, but he never tells Julia that he loves her. He loves his boy, though, and having a son has changed his attitude about things. He knows what it's like growing up without parents and feels a real responsibility to care for the child.'

Test Type: **Paternity**
Test Result: **Positive**

CASE STUDY: VIC, BIRMINGHAM

For two years, Vic Barker wondered if his son was actually his. He had spent three years with his partner and during that time had had a son with her. He was the only one in the relationship who worked, so his income alone cared for the child. He loved his son but his relationship with the mother was put under a lot of strain, and this caused him to worry about what should be done.

Roughly a year after their son was born, Vic and his partner decided to split. Soon afterwards, he discovered that she'd been sleeping around during their time together. When he confronted her on the matter, she seemed unsure and approached the subject awkwardly. Thoughts about his relationship with his son started to enter Vic's mind and he found that he was constantly asking questions of others, such as, 'Does he look like me? Are we the same?'

These questions and others compelled him to have a DNA test taken. Although he feared the worst, the test results came back positive; he was the child's father. Needless to say, Vic was very happy about it, although if the results had been negative it would have been easy for him to cut ties, even though he wouldn't have wanted to do this. Today, although his relationship with the mother isn't going too well, he does all he can to be with his son.

Regarding his experience, Vic says, 'If you've got any doubts, DNA testing is the best way to go. There comes a point in all the doubt and questions when you start giving up. Having the results gives you direction into life and helps you know what to do. I don't care how hard it is; I'll try to do all I can to raise my son.'

Test Type: **Paternity**
Test Result: **Negative**

CASE STUDY: JANE, BRISTOL

After being engaged in an eighteen-month on/off relationship with Reis, Jane was growing tired. In all that time, they'd really been dating for only about three months. He would get bored or angry and disappear for months at a time and then would return, asking for forgiveness. She wanted to move and find something more stable and less dramatic.

Shortly after breaking things off with Reis, Jane met up with Trevor, an old friend she'd met at school, on the rebound. Their time together was brief and never coalesced into a relationship, and they both moved quickly on to other things. However, it wasn't long after her split from Reis and her time with Trevor that Jane discovered that she was pregnant. According to the doctors, the baby had been conceived while she'd been with Reis.

When she first informed Reis that she was pregnant and that he was the father, he seemed to take it quite well, but it wasn't long before he stopped returning phone calls and cut off all contact with Jane. Only a few months before the birth, however, he called Jane and told her that he wanted to see the baby and be a part of its life. A week after the little girl was born, he came to visit them, but then he didn't show up until six months later. Jane was frustrated by his lack of interest and infuriated by his unwillingness to pay maintenance. She started to suspect that the reason why Reis was keeping his distance was because he doubted that he was really the baby's father. She considered getting a DNA test taken to prove to him that he was, but she was worried that it might be difficult to contact him, judging by his past behaviour.

It later transpired that Jane needed to get a passport for her baby and needed to get a DNA test taken in order to obtain the necessary documentation. This prompted her to get in touch with Reis, who was willing to undergo testing. The results were sent to Jane's and Reis's houses, and Jane fully expected them to come back positive.

They didn't. The test was negative, and Jane was stunned. Who *was* the father of her little boy? She had no idea what to think. The only other person it could have been was Trevor, but she'd been with him weeks after she had apparently conceived. Could the doctors have been that far out in their estimation of a conception date? Although she was shocked and surprised, Jane knew that Reis was most likely relieved, and indeed she never heard from him after the results had come through.

Jane just had to accept that the doctors had made a mistake and approach Trevor to let him know that he had a daughter. It had been over a year since they'd last spoken and Jane wondered how he would react to the news, but she needn't have worried: 'He was great and agreed straight away to have a test done.' Jane was happy, too, that he was willing to know whether or not he was the father.

When the test came back positive, Trevor responded well and offered to help to support the baby, and Jane was relieved that this potentially awkward situation was turning out to be so agreeable. She was doubly thankful to know that Trevor was a friend and someone that she had known for a long time. His reaction was supportive and helpful, and it brought a renewed calmness to what had previously been a stressful situation.

'I'm glad it worked out as it did,' says Jane. 'Reis was never interested in this baby, so he never created a bond. No one has missed out here.' Regarding her situation, she says, 'It feels good to share my story with others. And at the end of the day, every child needs to know who their father is.'

Test Type: **Paternity**
Test Result: **Positive**

CASE STUDY: REBECCA, PLYMOUTH

Rebecca had been seeing Simon for a few years, and when their relationship came to an end she got together with Matthew. After only a month with Matthew, she fell pregnant and was sure that the baby belonged to Simon, so she got back with him and tried to build a family. Simon, however, wasn't interested in having a child and left them both not long after the child was born.

Discouraged but ready to move on, Rebecca and Matthew renewed their relationship. They got on very well together and Matthew always treated Rebecca's son as if he was his own. As the years passed, however, Rebecca always held doubts in the back of her mind. What would she tell her son when he started to notice that he didn't look like Matthew? What would happen if she and Matthew decided to split up? She decided that she wanted to take a DNA test for her own peace of mind. She had no intention of telling Simon about her plan, though; he'd walked out nearly five years ago and didn't deserve a place in his son's life.

Knowing that taking the test was something she had to do was one thing; finding out how to go about doing it was another. Rebecca was short on money and worried that the legal processes and fees would be more than she could bear. She also felt really guilty that she was involving Matthew and her son in her mess in the first place. She felt lost and didn't know what to do. Then one day she happened to be looking in a magazine and saw an ad for a DNA-testing company that also featured a story of a woman in a similar situation to hers who had used their services, and Rebecca felt convinced that this was something she could also do. It was no longer a matter of waiting; the possibility had just become a reality.

Because she didn't want Simon to know that she was having the test taken, she asked Matthew to do the test with her, on the assumption that the negative results from this test would prove by elimination Simon's biological relationship with her son. She was convinced that Simon was the father and never thought it would be anyone else. She just wanted to know for sure.

When the results were delivered, Rebecca opened them alone. To her shock and surprise, they were positive. Simon really wasn't the father after all; Matthew was. Bewildered, Rebecca didn't know what to do. She feared that telling Matthew would make him angry, but she knew that she had to let him know, so she phoned him at work and prepared for the worst. Again, she was met with surprise: Matthew was overjoyed. He already loved the child as his own son, and now he'd found out that the boy he loved *was* his own son. Rebecca was understandably relieved.

As the days passed, Rebecca started to notice that there *were* similarities between her son and Matthew. When they shared the good news with their friends and families, several of them admitted to having believed that Matthew was the father the whole time. Even Matthew considered the possibility. Rebecca was astonished, as she'd thought that they had all agreed with her that Simon was the father. It was all too strange, but very lovely. Rebecca and Matthew are, needless to say, very happy to be a real family at last.

Test Type: **Paternity**
Test Result: **Positive**

CASE STUDY: ALEX, HULL

For five years, Alex had led the life of a married man. His wife had produced two young children, a boy and a daughter, now four years and nearly eighteen months old, respectively. Like most fathers, Alex had never questioned the fact that the children were his, especially his son, who resembled him greatly. It wasn't until he found out that his wife had been having an affair for most of their marriage that he started to have doubts.

Fearing that his wife might deny him any and all contact with the children, Alex never told her of his suspicions, but he considered the fearful possibility that his children might not be his after all.

Among the lies and deceptions, Alex felt that he'd been caught up in something that perhaps wasn't what it seemed. He found it hard to look at his children, especially his daughter, who didn't look much like him, and resolved that he would have their DNA tested in order to vanquish his doubts and give him, for once, the truth he was searching for.

As Alex waited for the right time to carry out the test, he felt increasingly strongly that this was the right thing to do. After all, if his wife had lied about so many things, there was every possibility that she would lie about the legitimacy of her daughter (about whose paternity Alex had the strongest suspicions), and Alex didn't want to raise and care for a child who wasn't his.

On the other hand, Alex loved his children and hoped, despite the hurt and betrayal he felt, that the test results would prove that he was the father of them both, and so he sent off samples to be tested.

One week later, the test came back with positive results for both children. 'Receiving the results was the full stop I needed for that time of my life,' Alex reports. 'I'm 110 per cent glad that I had [it] done. I can't say enough about the peace of mind it brought.'

Test Type: **Paternity**
Test Result: **Positive**

CASE STUDY: ANNA

When she'd been growing up, Anna had wondered why she looked so different to the rest of her family, especially her older sister, who was five years her senior. Her mother used to joke that she'd found Anna under a cabbage leaf, and it wasn't until she was twenty-five years old that Anna discovered the real reason.

When she was eleven years old, Anna's parents separated. Although the family lived in a small town, Anna didn't have much contact with her father after that; nor was she very close to him. Five years later, her mother married an old family friend whom Anna had known her entire life. She was quite close to him, as she'd known him for so long, and was happy that her mother had found someone that everyone knew and liked so well.

For the next five years, everything seemed to go well, until Anna's mother started to hint at the fact that the man Anna knew as her father might not have been her actual biological father. As the story began to unwind, Anna came to realise that the family friend her mother had remarried was actually a man her mother had had an affair with some twenty-two years earlier.

When Anna asked her mother why she'd waited so long to tell her, she was met with answers that reflected genuine worry concerning how her daughter would respond to the news. Her mother feared that Anna would lose respect for her after she knew what she'd done while she was still married.

Despite her initial shock and disbelief at the possibility of having a different father from the man she'd called Dad, Anna was curious and determined to find out the truth, so she sent some samples to a DNA-testing company. She waited eagerly for the results, hoping that they would confirm that she was actually the daughter of her mother's new husband, and it gave her comfort to see how they had the same nose, the same green eyes and the same easy-going attitude, so different from the rest of her family. She was also very close to her mother's new husband — so close, in fact, that he'd always seemed more like a father to her than the man she'd always thought of as being her dad.

When she received the results, Anna was elated to discover that her suspicions were confirmed: she was the product of her mother's affair twenty-two years earlier. She finally knew who she was and felt a connection with the man she now knew as her biological father that she'd never felt with anyone else.

Anna admits that the happy ending to her story depended largely on her ability to accept the results and that they came to her at a time when she was willing and ready to know about who she was. Had the results been different, her peace of mind and concern over her identity would have been greatly affected. However, she is happy to share her success story with others because it dramatically changed her life for the better.

Test Type: **Paternity**
Test Result: **Negative**

CASE STUDY: CURTIS, LONDON

Life as a teenager can be hard. Everything comes at you fast, and adjusting to new responsibilities can sometimes feel overwhelming. Seventeen-year-old Curtis Marrow certainly found that this was true of his life. What made it even more complicated, however, was the fact that his girlfriend told him she was pregnant.

Curtis had known the girl for many years, and their relationship had never been more than friends, except for one night that they shared together several months prior to this. When she found out that she was expecting, she was convinced that Curtis was the father, even though Curtis knew that she'd been with several other young men around that time.

Throughout the pregnancy, the girl persisted to claim that she was 100 per cent sure that Curtis was the father. Curtis's family knew her and felt inclined to believe her. In fact, *everyone* believed her. Curtis was young, too young to be a father, but his friends and family were very supportive of him and promised to help him out with all his responsibilities.

When the baby was born, it had dark hair and dark eyes, just like Curtis's. No one really disputed that he was the father, and his relatives had grown excited about the prospect of having a baby in the family. Mothers were now grandmothers, brothers and sisters were aunts and uncles and Curtis found himself in the new role of father. They all bonded with the baby and felt very close to the new child.

Still, though, the question remained: was the boy really Curtis's son? Curtis and his family decided to get a DNA test done to make sure. They sent away some samples and waited.

When the test results were delivered, they

turned out to be negative. Curtis's family cheered with joy at finally knowing the truth but, when his mother, Angela, took them over to the girl's house, the new mother was devastated and apologised profusely, declaring that she'd honestly believed that Curtis had been the father. Meanwhile, Curtis and his family regretted the fact that her lovely baby boy wouldn't be a part of their lives any longer. Nevertheless, the negative result meant a lot to Curtis, as it allowed him to move on with his life instead of having the responsibilities of fatherhood thrust upon his shoulders at a mere seventeen years of age.

And life has moved on. 'There was no falling out over anything,' says Angela. 'I still see the baby every once in a while, just in passing on the street.' She admits that she was sad not to be a grandma but happy for her son.

Everyone in Curtis's family hopes that the girl finds out who the father is. 'There are so many kids that grow up without ever knowing,' Angela notes. 'We hope that, for the sake of this child, it never has to grow up wondering.'

Test Type: **Paternity**
Test Result: **Positive**

CASE STUDY: SEAN, MANSFIELD

Sean had been with his partner for a year before they decided to split. Following the break-up, several months passed, and then she contacted him to tell him that she was pregnant and that the baby was his. Sean asked her how this could be the case when it had been so many months ago that they were together. It was possible, she informed him, because the baby was due in less than two months.

Sean didn't know what to think. He was only nineteen and didn't feel ready to take on the responsibilities of fatherhood. He had certainly never planned to start a family at this time of his life. He also mistrusted this girl and her claims: why had she waited so long to inform him? Either way, he felt that something should be done, but privately and quickly so that he could just be at peace with himself.

Two months after his daughter was born, Sean had her and his DNA tested to determine if he was her father. When the results came back positive, Sean knew that he had a choice to make: he could either come to terms with things and support the baby or he could pack his bags and move to America. Despite the appeal of the latter choice, Sean was surprisingly jubilant. 'I was over the moon when I found out,' he says. 'Now I can support her 100 per cent.' He and his daughter's mother aren't together, but she is nonetheless thankful for Sean's support, while Sean is happy to know that he has a daughter and enjoys the time he gets to spend with her.

Test Type: **Paternity**
Test Result: **Positive**

CASE STUDY: MELISSA, HERTFORDSHIRE

Melissa conceived and bore a small daughter with her husband, Darryl, but only eighteen months after Moira was born, Darryl and Melissa divorced. Darryl remarried sometime later and had three children with his new wife, whereas Melissa remained a single mother.

For Moira's sake, Darryl and Melissa maintained contact until their daughter was four and a half years old, at which point Moira decided that she didn't want to have any further contact with her father. Thereafter, Darryl and Melissa lost touch with each other.

Then one Christmas, nearly ten years later,

Moira decided that she wanted to see her father again and asked her mother to put them in contact. Reunited, Moira and Darryl quickly bonded and became quite close, much to the dismay of Darryl's new wife.

Perceiving Moira as a threat to her family, Darryl's second wife began telling lies and spreading vicious rumours about the legitimacy of Moira's biological relationship with her father, stating that Melissa and Darryl had been living a lie and that Moira shouldn't be allowed to mix with his new family. Although there were never any doubts between Melissa and Darryl, a seed of disbelief had been sown in Moira and the rumours were making her distressed and confused. And so, both for her own peace of mind and to put an end to the claims of Darryl's second wife, Moira asked her parents if they might have a DNA test performed to prove that she truly was the daughter of Darryl.

In the spring of 2005, Darryl and Moira were tested and were glad to discover that they were, indeed, father and daughter. During this time, Darryl had split from his second wife and began a renewed relationship with Melissa.

Darryl's second wife then began legal proceedings and refused to believe the results of the test, claiming it to be a hoax concocted by Darryl's first family. Despite this, however, Melissa and Darryl have grown closer and happier as a family.

Since that time, the family has become even closer. Moira has found in her father someone new to connect with, and she finds joy in noticing the similarities she shares with him, a joy she had always experienced but which grew stronger with the knowledge of the fact that he was, indeed, her biological father. Melissa says that, now they're all together, it's hard to tell that there had ever been a separation.

'The hardest part about doing the test was feeling forced to do it,' Melissa recalls. 'We'd never had any doubts or qualms that Moira was our daughter, but to have a DNA test is initially very daunting, because you doubt what you know in your own mind. Your automatic reaction is to question yourself.' For Melissa, those seven days between sending off the samples and receiving the test results were the worst days of their lives, but they found comfort in reading the test results together, and found reason to rejoice when the results confirmed what they already knew to be true.

'Moira is a very fragile girl,' says Melissa, 'but she is also very grown up and has been very mature about all this.' Moira has even explained the DNA-testing process to her friends and younger brothers and sisters. Hers has been a unique experience that has created a unique bond with her parents.

'At the end of the day, you're all very experienced to know,' says Melissa, regarding the situation they were faced with, which she believes strengthened the family. 'Having gone through this process, we know we can get through anything!'

Test Type: **Paternity**
Test Result: **Positive**

CASE STUDY: TOM, LONDON

Tom had a brief relationship with a woman named Angela. After a couple of months, their time together ended and they both moved on with their lives.

Five years later, Angela called Tom and gave him some shocking news. Sometime during their time together, she had fallen pregnant. She hadn't contacted Tom until this point because, perhaps, she wasn't sure herself who the father was. Now, five years down the road, her daughter was starting to ask questions that Angela was unable to answer. The fact was that, after Tom and Angela had ended their relationship, Angela took up with her previous partner. They were together for some time, but things ended badly when, faced with her daughter's queries, Angela had asked this man to take a DNA test. She was surprised to find out that the child was not his, and thus was forced to contact Tom.

Life had changed for Tom, too. He was now in a new relationship and had a son of his own from a previous relationship. The news that he now might have a daughter caused him considerable stress, and he knew that, if it was true, it would change his life completely. His present relationship was still fairly new, and the chance of adding a five-year-old daughter to the equation – a daughter he'd never known – complicated things even further. Yet, despite it all, Tom and Angela discussed possible courses of action and decided that they would have another DNA test done to see if Tom was the father.

Tom confessed that he didn't know what the test would prove. He was still shocked by the situation in which he now found himself. He worried that the girl was his and that, if she was, the news would damage his relationship with his new partner and would affect his relationship with his son.

When the results came through, they were positive. Tom discovered that he did indeed have another child.

Tom admits that adapting to the new situation has been difficult. 'Trying to bond with [my daughter] is difficult because she's older,' he admits. Yet there have been small blessings, too; his new partner has been very supportive, and his son has accepted the fact that he has a new little sister. Despite the complexity of the situation, Tom remains upbeat: 'We seem to be quite happy, and we try to find time to get together.'

Test Type: **Paternity**
Test Result: **Positive**

CASE STUDY: SAM, GLASGOW
Sam had had many on/off relationships, and his

time with Sharon wasn't that different. They regularly got together but didn't have a very solid relationship with one another. They eventually ended their time together, but a few weeks later Sharon called him up and told him that she was pregnant.

Due to the brevity of their relationship and their tendency to spend time with other people, Sam felt many concerns and doubts regarding the validity of Sharon's claim that he was the father of her child. His disbelief was met with so many accusations and so much anger that he decided to drop the case and prepared to care for the baby.

For a year, Sam was preoccupied with his doubt. Was the child his? And what would he do if it wasn't? Would he carry on as if nothing was wrong? Would he raise the child and support it as best he could? He felt torn and constantly asked himself, 'How can I give 100 per cent care if I'm not 100 per cent sure this child is mine?' He felt uneasy and anxious about what he should do.

Sam's apprehension was only heightened by the fact that many of his friends and relatives misunderstood his intention to perform a DNA test on the child. They didn't understand why he wanted to know and felt that he should just assume responsibility and care for the child. Furthermore, knowing that it was an important decision, he struggled with the task of finding a testing company he could trust to give him the valid and legitimate results he so desperately sought.

Despite his fears and anxiety, however, Sam had to know the truth. No matter how much the child looked like him, no matter what he thought or even wanted to believe, and no matter what anyone else said, Sam knew that he would never be satisfied until he knew for certain that he was the biological father of this child.

A year after the child's birth, Sam sent off for a testing kit and carried out the procedure. He decided not to say anything to the mother, as he'd initially met a great deal of resistance from her when he suggested taking the test. He felt that the test was only for his own peace of mind, after all. If the test did turn out to be positive, he could put all his heart into caring for the child.

A week later, the test results were delivered, and Sam was relieved to find out that he was in fact the child's biological father. He felt as though a huge weight had been lifted from his shoulders. The child, a son, is now over three years old and, even though he and Sharon aren't together, Sam spends time with him on a regular basis. He enjoys being with his son and happily promises, 'That will always be the case.'

Test Type: **Paternity**
Test Result: **Negative**

CASE STUDY: JEREMY, HERTFORDSHIRE

While Jeremy was married, he and his wife were rarely intimate and they drifted apart. Then one night while they were out, his wife confessed that her love for him was strictly platonic and nothing else. With that news, combined with various other factors, they decided that they should both move on to find real love.

Barely a week later, Jeremy's wife called, apologising for telling him that she didn't love him and telling him that it had been a mistake for them to split. In the end, they decided that getting back together would be better than divorcing, so they gave their relationship one more try.

Two weeks after they'd reunited, Jeremy's wife told him that she was pregnant. Knowing how infrequent their intimacy had been, he found it strange that she would fall pregnant at such a volatile time in their relationship. He also thought that, if she had indeed fallen pregnant with his child, it wasn't necessarily the way to fix a marriage. However, throughout the pregnancy and even three years after the birth of their son, Jeremy and his wife stayed together.

Not long after those three years, Jeremy and his wife split again, this time for good, and soon Jeremy's doubts resurfaced and he found that he couldn't help but doubt his paternity of his son. At this point, Jeremy was young and poor, with no way to fund his suspicion, and he decided to put the issue on the back burner for now. After all, he loved his son and didn't want to abandon him. The test would have been solely for his own peace of mind.

During the next seven years, Jeremy's ex-wife used their son as a tool with which to manipulate and emotionally blackmail Jeremy, who, frustrated, finally determined to get a DNA test taken to settle the issue once and for all. By this time he'd started his own business and had become quite successful. He also had a new partner and a four-year-old daughter to dote upon. He decided that he had to do something in order to prepare for the future, and now seemed as good a time as any.

Jeremy isn't the type of person to rush into things. He knew that getting a DNA test performed could have serious consequences and wanted to make sure that he was fully prepared to accept what he found out to be the truth. He spoke with many of his close family members and friends. What would they do in his situation? Was taking a DNA test the right way to go, or should he just let sleeping dogs lie?

Jeremy decided to get the test done without telling his ex-wife, but when the results came back negative he was both shocked and saddened, as he'd wanted to know that the child was actually his. He jumped into his car and took a long drive. He wanted to get advice, take it in, digest it and come to terms with things before he acted on the information he'd received. He turned over a series of questions and reflected about what he should do and how – or, indeed, whether – he should break

159

the news to his ex-wife. He kept thinking about all the times she'd used their son as a pawn in her battles with him and felt vindicated to know that, for once, he held all the aces. He also knew that, if ever her son wanted to know who his father was, she wouldn't know what to say. Jeremy didn't want her son to feel like any of this had been his fault, so he resolved that he would never tell him – or, at least, not until he was much older. He couldn't bear to think that he didn't have a son, and he didn't want his son to be without a father.

Several months after Jeremy had the test done, he has yet to tell his ex-wife of the results. And what does he expect her to say when he tells her? 'She could turn around and be nasty, but I hope she'll be able to look at her son's interests and see how it would affect him,' he says. He'll likewise make an agreement to keep paying a minimum maintenance fee, but nothing extra, especially nothing that would finance her (and her partner's) standard of living. He's doing this only for his son.

'I haven't come to terms with [the results] yet,' Jeremy confides. '[My current partner and

I] enjoy having him over, and he's got a very stable life when he comes to stay with us.'

All in all, Jeremy is happy with the peace of mind he has achieved from having the test done. He looks forward to the day when he can get the rest of the situation figured out. 'Time is a good way to get things right in your head.'

Test Type: **Paternity**
Test Result: **Positive**

CASE STUDY: AARON, MANCHESTER

Aaron and Melanie recently ended a two-and-a-half-year-long sporadic relationship. 'We just didn't get along that well,' he says, adding that they were always leaving each other and getting back together. It was a very draining process.

During those years, the couple had a son together. Then, after they'd split up, Aaron heard rumours through his friends that, while he and Melanie had been together, she'd been cheating on him with someone else. Although she denied this at first, she ended up confirming it later on.

Feeling betrayed and angered by the news, Aaron started to wonder if the boy he'd thought was his own could be someone else's. He wanted to have things clear in his mind because he really wanted to act wholeheartedly towards his son. He was likewise eager to avoid any legal proceedings that could be anticipated in the future. However, he feared approaching Melanie because he didn't want her to mistrust him even more than she already did, and deep down he truly felt guilty for even thinking about having his son tested. After weighing up all the pros and cons regarding the situation, though, he wasted no time in ordering and administering the test.

By the time the test results were delivered, Aaron found that he wasn't feeling so unsure about his son's paternity. He'd been noticing points of similarity in their looks and felt strongly that the test would come back positive – which, to his relief, it did.

'Having the test done has only made me more committed to my son,' affirms Aaron, who sees his son every weekend, although his work takes him away most weekdays. 'I love coming back to see him,' he says. 'It's a good thing now.'

Test Type: **Paternity**
Test Result: **Positive**

CASE STUDY: REBECCA, LONDON

When Rebecca gave birth to a beautiful little girl, she called up her ex-boyfriend and asked him to come with her and sign the birth certificate, but she was shocked and appalled when he refused, claiming that the baby wasn't his, especially as she'd already had two children by him and the third looked exactly like him. Nevertheless, her ex-boyfriend refused to take responsibility for fathering Rebecca's daughter and wouldn't sign the document.

Rebecca's relationship with the father of her two older children had been a little rough in the past. It was true that he was a good father, but really only when it was convenient for him. He and Rebecca had had a constantly on/off relationship that eventually ended when she found him in their bed with the family's babysitter. It wasn't long after that that Rebecca found out that she, herself, was pregnant.

During her pregnancy, Rebecca was stressed and angry. She was already accustomed to being effectively a single mother, but the prospect of raising one more child on her own was at times unbearable. Had she known that DNA testing was possible while the baby was in her womb, she would have had one done to eliminate the grief and anxiety she was feeling.

After the birth of the child and her boyfriend's reaction, Rebecca felt even more stress. With her boyfriend's constant denial and the babysitter issuing threats of physical violence against her, Rebecca felt alone and scared.

Four weeks after the birth, Rebecca decided to get to the bottom of this issue and persuaded her ex-boyfriend to take a paternity test. When the results arrived a week later, it was a positive match. Relieved and feeling validated, Rebecca sent the results to the father, who responded with anger and disbelief, ripping up the letter and claiming that the results were false. However, the proof of the test is undeniable and he is now required to provide financial relief to the family.

'Had I been younger, I might not have been able to cope with [the situation],' confesses Rebecca now. 'Younger mums don't have many people to turn to. This testing shows them that they have options.'

Rebecca couldn't be happier with the test results. She suffered a great deal of stress throughout her ordeal, and having the DNA test done gave her long-awaited peace of mind. She is relieved to have three beautiful children, and especially grateful for the new bundle of joy that has entered the home.

Test Type: **Paternity**
Test Result: **Negative**

CASE STUDY: RONNIE, BEDFORDSHIRE

Ronnie and Elaine had been together for nearly six years, but their work situations kept them apart so often that they managed to get together only two or three times each year. Due to the difficulties and strain that the long distance imposed on their relationship, they decided to split up, but in a moment of weakness they got back together again for a night.

A few months later, Elaine contacted Ronnie to tell him that she was pregnant and that she'd conceived near the time they'd been together. She also admitted to seeing another man during that time but couldn't bring herself to tell the other man that she'd cheated on him with Ronnie. Although it was possible that Ronnie was the father, he was suspicious and Elaine was unsure.

Ronnie knew that something had to be done, as he'd been planning to go to South Africa for a year to work and Elaine's pregnancy could put those plans on hold indefinitely. He found a DNA-testing company's website on the internet and decided to take the test.

As he waited for the results, he hoped that the baby would turn out not to be his and that he could carry through with his plans. As it turned out, the test results came back negative, much to Ronnie's relief. He could now move on with his life. Nevertheless, he is still friends with Elaine, who now knows that the other man is the father of her child. She is likewise glad to know the truth and happy that things turned out the way they did.

Test Type: **Paternity**
Test Result: **Negative**

CASE STUDY: PAUL, NORTHAMPTON

Paul and Rosie had been together for only six weeks when she told him that she was pregnant, although she confessed that she couldn't confirm that Paul was the father. As he had grown up without a father, Paul vowed that he would care for this child.

During the next six years, Paul lived in a state of doubt, unsure whether or not the boy was actually his son, and consequently never really forged a real bond with him, notwithstanding the contact he maintained with him and Rosie over the years.

Paul admits that, in his heart, he never really felt that the child was his, but remarks made by others commenting on how similar they looked caused him to feel that perhaps the child really did belong to him.

To find out once and for all, Paul and Rosie both agreed to have a paternity test done, and to their surprise (and Paul's relief) the test came back negative: the child wasn't his. Paul now admits that the test served only to confirm what he'd always thought, but that he felt a responsibility to take care of the child, as he knew what it was like to grow up without knowing one's father.

Rosie encouraged Paul to reconsider the results, hoping to rekindle their relationship, and even doubted their accuracy, telling Paul that she'd never been with someone else while they'd been together. Since the test, however, Paul hasn't seen much of her and her son. 'Now I can move forward with my life without feeling guilty,' says Paul, 'and it's become more important to me to have a child with someone I really love.'

Test Type: **Paternity**
Test Result: **Negative**

CASE STUDY: ASHLEY, UK

Ashley has a unique and long story of love. A devout Muslim, at the age of twenty-six he'd been part of an arranged marriage. Having never loved his wife, the marriage was never consummated and was eventually annulled only three weeks later, against the wishes of his parents. Ashley decided to look for love on his own terms and he looked forward to meeting a woman he could spend the rest of his life with.

As the years passed, Ashley became a college professor, and during one semester he found that he connected well with Judith, one of his students. The two of them decided to pursue a relationship at the end of that semester.

Ashley's parents were predictably wary of the

relationship, as Judith was a Catholic and they'd hoped that Ashley would marry a Muslim girl, although they couldn't deny that Ashley seemed truly happy to be with Judith. She and Ashley stayed together for nearly seven years and looked forward to having their own home and life together.

Late in 2002, Ashley visited Mecca, and when he returned he was surprised and hurt to learn that Judith no longer wanted to see him. As he struggled through the next few months, he wondered what had happened and what he might have done to have upset her. Then, at the beginning of the following year, Judith called him to tell him that she was pregnant and that he was the father.

In March 2003, Judith and Ashley got back together and eagerly anticipated the birth of their baby daughter that August. When Jemma was born, Ashley in particular struck up a close bond with his daughter. For the next year, all seemed to be going well.

Then, in the summer of 2004, Ashley and Judith's relationship deteriorated. Judith often spent a lot of time away from the house and the family, claiming that she was depressed and needed to get away. Ashley never had reason to doubt or mistrust Judith's claims and worried about her constantly. He felt that if they could get a house of their own then perhaps she would feel more like part of the family and would feel better about spending more time with him and baby Jemma, so that's exactly what he did. However, this didn't solve the problem. In fact, things got worse.

By chance, Ashley met up with an old friend, Tariq, whom he hadn't seen for quite some time. He told him all about his daughter and even showed her a picture. After just one look at the picture, Tariq, a doctor of biomedics, told Ashley that the young girl definitely wasn't his daughter. 'Tariq said he could just tell by looking at the picture and knowing about DNA,' recalls Ashley. In fact, the most obvious factor supporting Tariq's claim was that, while Ashley was Asian/Indian and Judith was Caucasian, Jemma had no Asian/Indian look to her at all.

Having trusted Judith, Ashley had never questioned the fact that his daughter really looked anything like him. Now, confused, he confronted Judith, who admitted that he was most probably not Jemma's father. She further told Ashley that, during all the time she'd spent away, she'd actually been with Jemma's biological father, who apparently wanted to be a part of Judith's life again. She told Ashley that it would be better if he broke off contact with Jemma until things could get sorted out. It was during this time that Ashley decided to get a DNA test taken, just to confirm what he hoped wasn't true.

Ashley had the test done in January 2005, and as he waited for the results his thoughts were split two ways. One thought was to just let it be; he'd always felt that destiny was destiny and that things could not be changed. His other, more hopeful thought was that, even if the test results came back negative, he could still have contact and, essentially, a relationship with his daughter. The results, however, were as he feared, negative, and all contact with his daughter ceased. 'I've missed Jemma since the end of January,' he mourns. 'I raised that baby for two years. I couldn't bear to say goodbye... I am susceptible and naive, [and] I find it hard to trust anyone these days.'

Despite his grief, Ashley took the opportunity to go to India to help with the relief operation after the tsunami that struck at the end of 2005, and, while he was there, he witnessed a lot of death, poverty and sickness – things that gave him a great perspective on life. He feels now that there's a lot he can do for others and acknowledges that, for the time being, this is some solace.

But what does Ashley's future hold? 'I can only take one day at a time,' he says. 'This has been a very traumatic year for me. I'm still single, but I want children.'

Test Type: **Paternity**
Test Result: **Negative**

CASE STUDY: TIM, NOTTINGHAM

At twenty-five years old, Tim seemed to have a virtually carefree life. He was young, studying at university and going out with a girl with whom he thought he was in love. But when his girlfriend called him at college one day to tell him she was pregnant, it seemed to him that his life was about to get more complicated.

The news was not only shocking but distressing, too. Tim felt that he and his girlfriend were just kids themselves and didn't feel prepared in any way to care for a new child. Frequent arguments disrupted their once loving relationship, which eventually dissolved.

Despite the break-up, Tim and his girlfriend managed to stay civil with each other and Tim resolved that he would not step back from this situation because the baby his girlfriend was carrying was, after all, his child.

After the birth of their daughter, Tim and his ex-girlfriend proceeded down the legal route and prepared to have some routine lab work done. One day, during a meeting at the hospital, a social worker advised Tim that he shouldn't get too attached to the child because she might not really be his.

Confusion began to creep in, and, although Tim and his girlfriend were no longer together, he still attended the christening of their daughter. 'During the christening of the child, I felt really awkward,' recalls Tim, 'like, "Is she or isn't she mine?"' His doubts encouraged him to get the child's DNA tested.

After searching on the internet, Tim found a DNA-testing company that would provide him the peace of mind he was seeking and he contacted them. As he waited for his test results to come back, he prepared himself for either outcome. When they came back negative, he was extremely relieved. Had they come back positive, he would have been happy to have a daughter, but the negative result meant that he could move forward with his life.

Since finding out he wasn't the child's father, Tim has essentially yet amicably cut all ties with the mother and daughter and is relieved to have some peace of mind regarding the matter. Working with the DNA-testing company was, for him, a very positive experience: 'Dealing with the company was brilliant. There were days that I would call three of four times, trying to figure out what to do and what things meant. Everyone was so patient and helpful.'

Test Type: **Paternity**
Test Result: **Positive**

CASE STUDY: ROB, ISLE OF WIGHT

Rob was in a sticky situation. His partner, Lisa, had been pushing him to his limits. For two years he had become used to her Dr Jekyll and Mrs Hyde-type personality, but he was coming to mistrust anything and everything she told him.

So, when she told him that she was pregnant, he thought it was another one of her lies.

Rob and Lisa had been together for about twelve months when they first split up. After a few months, they got back together and Lisa fell pregnant right away – or, at least, that's what she told Rob. He'd learned to be sceptical of her claims and his suspicions arose when Lisa insisted that her baby would be born at least five weeks premature. Rob didn't know what to think, as he could only take Lisa's word for it, but he wondered if this prediction was a cover for something she'd done while they were on hiatus for those few months.

Rob and Lisa stayed together for only a few more months into her pregnancy, after which Rob decided that he was through with her petty behaviour. 'I'm the kind of person that is open to discussing things,' he says. 'Everything felt so tedious and I was tired of it.' He wanted peace of mind about the matter and considered telling Lisa flat out that he thought she was lying, but he knew it would only make her even more upset and angry (she was already threatening to withhold the baby from him, amongst other things). She had effectively shut Rob out.

As Lisa had predicted, the baby was eventually born five weeks prematurely, which only heightened Rob's desire to get a DNA-based paternity test done, and so, when the baby was old enough to receive the test (waiting helped to avoid unnecessary trauma to the premature infant), Rob asked Lisa if she would mind if he sent off some samples. She willingly agreed.

As he waited for the test results, Rob gauged what his reaction would be to each outcome. If the results were negative, he would be relieved not to have to pay child support and could end contact with Lisa. If they were positive, however, he would have a beautiful child and would have a say in its upbringing. He wrestled with what he hoped for, and the letter came back finally saying that there had been a positive match and that Rob was definitely the father. He felt a small satisfaction in knowing and was actually quite happy and glad the baby was his.

Rob now looked forward to taking a part in his child's life, but unfortunately it hasn't exactly panned out that way. 'I'm basically only in the picture because I provide the funding,' he says. '[Lisa] hardly lets me see the child, and she hasn't let me have any say in the child's upbringing... She has pushed me out of the whole process.' He admits to feeling rather frustrated. 'I can't do anything. I suffer the emotional abuse but I don't get the benefit of bonding with my child.'

Even though he has no plans to rekindle his relationship with Lisa, Rob has promised that he will remain faithful to the baby they had together. 'I've moved on from Lisa, but I could never turn my back on the baby.'

Test Type: **Siblingship**
Test Result: **Negative**

CASE STUDY: KERRY, YORKSHIRE

When fifty-nine-year-old Kerry complained of feeling pain in her joints and confessed to her half-brother that she suspected arthritis, he told her that it had to be gout, which ran in her family. But Kerry had no recollection of anyone in the family ever having gout and, when she insisted that her brother was wrong, he told her that he was, in fact, quite right: 'Gout runs in *your* family, with *your* father!' When Kerry showed absolute shock at his words, her half-brother realised that no one had ever told her who her father really was.

Kerry's half-brother's mother had a brother named Charles. Every summer the family went to Ireland to camp for a few weeks. By the time of Kerry's conception, her half-brother's parents had split but continued to gather together as a family for the holidays. According to her half-brother, Kerry's mother and Uncle Charles had had a summer fling and Kerry had been the result.

For Kerry, receiving this news fifty-nine years after the fact was a shock. Why had no one told her the truth? What did it mean? Who was her father? As she started to put the pieces together, she realised that the only way she might actually know for sure was through DNA testing.

Obviously, Kerry's half-brother wouldn't be able to help her with DNA testing, as it was now clear that they shared neither a father nor a mother. This left her to seek out family in Ireland and see if she could find another half-sibling through Uncle Charles that might prove that he was her father. The problem was that, at this point, many people who might have been able to tell her had died. It was going to prove a difficult task.

After months of searching, Kerry contacted one of Uncle Charles's daughters who she'd seen pictured in an old photograph with her two other sisters, who had all dispersed. When Kerry met her, she told her her reasons for seeking her out and explained that they might be half-sisters. Charles's daughter couldn't believe it – she was convinced that her father would never do anything of the sort – but she agreed to participate in a DNA test with Kerry so that Kerry might gain that peace of mind she was seeking. If the test gave a positive result, it would confirm what everyone in the family had thought this whole time: Kerry was the love child of her mother and Uncle Charles

To Kerry's family's surprise, the test results came back negative. Of course, this didn't automatically mean that Kerry's father had not been Uncle Charles, she could have been the product of a different affair, or she could actually be the daughter of her perceived father and mother, which would mean that a DNA test with her half-brother might yield useful results. They duly sent back samples, but when the results came back positive her half-brother denied their veracity, attesting that his father had always told him that Kerry wasn't his daughter.

Kerry, on the other hand, is not satisfied. 'I feel in a quandary,' she confesses. 'It would just be nice to know for sure.' She has planned further trips to Ireland to find more information concerning her heritage. She knows that she's missing something and is ready to get to the bottom of things: 'I am determined to find more proof!'

⑫

USEFUL CONTACTS

CSA (Child Support Agency)
The main office of the UK government dealing with matters of child support.

  Mainland UK office
  Ashdown House
  Sedlescombe Road North
  St Leonards on Sea
  East Sussex TN37 7NL
  Tel: 08457 133 133

  Northern Ireland office
  Great Northern Tower
  17 Great Victoria Street
  Belfast
  County Antrim BT2 7AP
  Tel: 0845 713 9896

DCA (Department of Constitutional Affairs)
Publishes guidelines on how to perform DNA testing
and offers a list of recommended companies.
   Selbourne House
   54 Victoria Street
   London SW1E 6QW
   Tel: 0207 210 8614

DWP (Department of Work and Pensions)
Responsible for delivering advice through the modern
network of services, including tax credit and child
benefit.

   Room 112
   The Adelphi
   1–11 John Adams Street
   London WC2N 6HT
   Tel: 020 7712 2171

Inland Revenue
Provides advice on tax issues regarding parents.

   Lyndhurst House
   120 Bunns Lane
   Mill Hill
   London NW7 2AP
   Tel: 020 8238 8600

New Deal For Lone Parents
A voluntary programme designed to help those lone
parents who are looking for work, are not in work or

work fewer than sixteen hours a week.
Tel: 0845 606 2626

Commission for Racial Equality
Specifically designed to promote racial equality.

St Dunstan House
201–211 Borough High Street
London SE1 1GZ
Tel: 020 7939 0000

Action for Prisoners' Families
A helpline offering advice and guidance to anyone
with a relative in prison anywhere in England or
Wales. Advice on visitation, childcare and keeping in
touch is freely available.
Tel: 0808 808 2003

Childline
Offers free and confidential advice to children and
young people via a twenty-four-hour helpline.
Provides counselling services throughout the UK.

London and the Southeast
Tel: 020 7650 3200

Wales
Tel: 0870 336 2935 (Swansea)
Tel: 0870 336 2930 (Rhyl)

Northern Ireland
Tel: 0289 032 7773 (Belfast)

Northwest
Tel: 0161 834 9945 (Manchester)
Tel: 0151 260 7590 (Liverpool)

Scotland
Tel: 0870 336 2910 (Glasgow)
Tel: 0870 336 2900 (Aberdeen)

Southwest
Tel: 0870 336 2905 (Newton Abbott)

Midlands
Tel: 0870 336 2915 (Birmingham)

Yorkshire and Northeast
Tel: 0113 244 4004 (Leeds)

Surestart

The UK government's programme to deliver the best start in life for every child by bringing together early education, childcare, health and family support.

Surestart Unit
Department of Education and Skills and the
Department for Work and Pensions
Level 2
Caxton House
Tothill Street
London SW1H 9NA
Tel: 0870 000 2288

Children's Legal Centre

Provides free advice and guidance to parents involved in disputes over education or access.

University of Essex

Wivenhoe Park
Colchester
Essex CO4 3SQ
Tel: 01206 873820

Children 1st

Provides support for families in difficulty in Scotland.

83 Whitehouse Loan
Edinburgh EH9 1AT
Tel: 0131 446 2300

Citizens' Advice Bureau

Provides free and independent advice on all matters, including finance, paternity testing and parental rights.

Myddelton House
115–123 Pentonville Road
London N1 9LZ
Tel: 020 7833 2181

Community Legal Advice Services

Provides guidance on all legal issues.
Tel: 0845 345 4345

Dads UK

Provides support and guidance in all matters, including access and finance.

85A Westbourne Street
Hove
East Sussex BN3 5PF
Tel: 01273 232997

Daycare Trust

A national UK childcare charity campaigning for quality affordable childcare.

21 St George's Street Road
London SE1 6ES
Tel: 020 7840 3350

Family Mediation Scotland

A voluntary organisation giving separating parents an amicable option when dealing with childcare issues.

18 York Place
Edinburgh EH1 3EP
Tel: 0131 558 9898

Families Need Fathers

A charitable organisation concerned with maintaining relationships between children and both parents.

134 Curtain Road
London EC24 3AR
Tel: 0870 760 7496

Family Rights Group
Offers help and support for those dealing with social services.

Tel: 0800 731 1696

Gingerbread
Offers support to lone parents in England and Wales
Tel: 0800 018 4318

Maternity Alliance
Offers support for both parents during pregnancy.

Third Floor West
2–6 Northburgh Street
London EC1V 0AY
Tel: 020 7490 7638

National Council for One-Parent Families
Operates an information service for lone parents and lobbies for changes in law to improve the current provisions for lone parent families.

255 Kentish Town Road
London NW5 2LX
Tel: 0800 018 5026

National Debtline
Provides advice for those experiencing debt in England, Scotland and Wales.

Tel: 0808 808 4000

National Council of Voluntary Childcare Organisations

Provides information of local voluntary childcare facilities across the UK.

Unit 4
Pride Court
80–82 White Lion Street
London N1 9PF
Tel: 020 7833 3319

National Family Mediation

A non-profit-making organisation offering mediation to families in England and Wales.

Alexander House
Telephone Avenue
Bristol BS1 4BS
Tel: 0117 904 2825

NSPCC (National Society for the Prevention of Cruelty to Children)

The UK's leading charity for preventing child cruelty.

Tel: 0808 800 5000 (helpline)

One-Parent Families Scotland

Offers representation for lone parents in Scotland.

13 Gayfields Square
Edinburgh EH1 3NH
Tel 0800 018 5026

Parents' Advice Centre (Northern Ireland)
Offers support for parents in Northern Ireland

Tel: 0808 8010 722

Parent Line Plus
Provides guidance for parents, step-parents,
grandparents and foster parents.

Tel: 0808 800 2222

Relate
Provides counselling for people with relationship
issues.

   Herbert Gray College
   Little Church Street
   Rugby
   Warwickshire CV21 3AP
   Tel: 0870 601 2121

Samaritans
Provide a twenty-four-hour helpline offering support
to those who are running out of options.

   Tel: 08457 909090

Shelter
Charity dedicated to improving the lives of homeless
people and those in poor accommodation.

88 Old Street
London EC1V 9HU
Tel: 0808 800 4444

Women's Aid
Promotes the protection of women and children who
are experiencing domestic violence.

Tel: 0808 2000 247

Shared Parenting Information Group
Promotes shared parenting after separation and
divorce.

278 Garraways
Wootten Bassett
Wiltshire SN4 8LL
Tel: 01793 851544